Proportioned Geometrics

Proportioned Geometrics

✦

The Unification of Space-Time Matter and Force

Frank Arca

iUniverse, Inc.

New York Lincoln Shanghai

Proportioned Geometrics
The Unification of Space-Time Matter and Force

iUniverse books may be ordered through booksellers or by contacting:

iUniverse
2021 Pine Lake Road, Suite 100
Lincoln, NE 68512
www.iuniverse.com
1-800-Authors (1-800-288-4677)

ISBN-13: 978-0-595-39017-5 (pbk)
ISBN-13: 978-0-595-83408-2 (ebk)
ISBN-10: 0-595-39017-X (pbk)
ISBN-10: 0-595-83408-6 (ebk)

Printed in the United States of America

Section I

The purpose of this paper is to introduce a likely solution to the origin of space, time, energy, and forces. These ideas compose the structure of our universe; a combination of matter influenced by the dynamics of force. However, where does space and time originate? Why does the universe take such a shape? How can all that is perceived and unperceived be unified so that at any one instant or place or idea, a simple equation arises, and that which is measured, the outcome is always equivalent.

An equation of such qualities, yields a ubiquitous balance, hence origins of the above may be derived universally. If one were to perceive nature thru the existence of an insect, one would derive the above equation, if one were any other unknown being out in the universe, one would derive the identical principles of such an equation. This unified equation contains a function for which there could be no further reducible functions. To illustrate, the proposed theory of the atomic particles contains laws and rules which govern such bodies at ultra microscopic scales, and inversely, the proposed theory of the planets, stars, galaxies, and the universe as a whole, contain as well verifiable functions and rules from which they are governed. For space-time, matter, and force to form a logical arrangement of both ranges, the physical laws would have to be summarized under one principle relation from which all proceeding laws begin.

If one asks, what is space and time and matter, then there ought to be a solution for that inquiry. If one asks, where do such things originate, and in what way is our own existence connected to such material formations; then there ought to exist an equation, an explanation, a single unifying framework which, for all times and places, yields the integration and differentiation of space, time, and energy. Something makes the universe set up in such a way that the margin for geometric manifestation opens, and therefore, the properties of physics produce an atmosphere for extraction of matter and force. What is energy without geometry? What scales the shape of space-time, and how does matter unify under one force?

The new theory of unification does not propose an absolute cause for the creation of all things, rather, the theory summarizes a geometrical connection that occurs not as the first cause, however, as the result of a greater source still unknown to our thoughts. What I am exploring is the geometry of the dynamics of matter entering the universe; how space-time, matter, and force become integrated to yield the order observed today. If one were to step outside of the mind and become a hinge of space-time, then this new force field would be overwhelmingly noticeable.

For those who do not believe in the physics and the scientific methods for deconstructing the world, I ask you to understand this theory not as a replacement for a creator, instead, if you would be kind enough to understand this theory as an aid to bring our thoughts nearer to the creator. I contest that faith is the strongest force in nature, and without it, there could be no science.

Please be assured that the contents of this volume do not propose the absolute cause of the creation of the universe, that this book does not suggest to disprove any notions concerning the existence of a God. I believe it is unwise to deduce that anything other than those of the powers defining the concept of God could create such a universe, and I am compelled to not question any individual's belief of such. I am a faithful person, and I believe that only God can dictate such claims. What I am questioning herein is the connection between that of the geometric Brain material and that of the fabric of space-time, consequently, I am inquiring as to the nature and significance of the Brain material as a non-biological system, more exactly, as a geometrical system influencing the forces and structure of space-time.

Proportioned Geometrics

The past laws of physics derived, and those still in progress, all innovate a partial solution to the exact and absolute way that unifies all ideas, matter, and space-time; however, the grand unit required to complete the equation of universal unification, is the Brain material. Ultimately, what ought to be discovered is, that at any and all points and events in space-time, and for all frames of reference, and that for any given system, that there would be a relationship, a function which therefore governs the properties of every physical structure in an equivalent dynamic.

The foundation for the unified field equation is based upon the principles of proportioned geometrics, that is, the framework of dynamics which arise from an immeasurable interaction of substance and vacuum. This property, that the Brain material interacts directly with the vacuum state and the fabric state of pre-matter, is the root for all forces, the electro-magnetic, the strong and weak forces, and gravitational. Matter is not a tangible quantity; the order in which it extracts into a real function is thru the dynamics of geometric proportions. The Brain is a substance which interacts with the vacuum, and at this point of interaction, space-time is extracted, and the pre-matter sheet is given to it the function of geometric proportionality with respect to the geometric Brain material, and this underlying principle sets off the equation of unification which can never be broken down.

Universal force: the Geometric Derivative of The Brain material and geometry. Exponential expansion, curvature, and universal force field, are all products of this interaction.

Property : If the material of the Brain enters the vacuum and under given conditions, then a function is induced, such that for all directions and points, there exists a field influenced and directly proportional to the geometrics of the Brain material.

Postulate one: The Brain material does not contain coordinates within the domain of space-time, therefore, this unit does not obey the same laws of physics as the matter which exists within space-time.

Theoretically, the Brain structure impresses an incalculable dent onto the zero space-time vacuum manifold, parallel to the zero pre-matter manifold. The distance between any object is zero, however, there are revised measurements for distances throughout the universe under the proposed geometry. The universal force, caused by that of the substance of the Brain's presence, and thus reorganizing a non-function manifold vacuum, into a dynamical system, yields and unifies the four current forces. Under this geometry, unit differentiation occurs, thus an equation for the initial behavior of the universe, and of the derivation of a function, from which all construction of physical laws by observation and by governing experimentations are obtained.

Proportioned geometrics deduces that the fundamental forces which influence the energy in space-time, are forces derived from one universal force field, more powerful than the sum of a trillion billion to the power of a trillion black holes, and even then, this still would not be close to the magnitude of power the Brain material contains first as a substance, and second as a substance with fields of force. This force formalizes an equation and induces a geometric function, a framework which constructs and deconstructs every observable and non observable unit of matter. The core of this equation, which from a reference of pure empirics, is the very material that has been in question for so long as the "grand phenomenon", thereby reducing the paradox of thought and perception, and directly relating space-time.

Section II

Pre-Matter

Pre-matter is the total points and total area of all matter, without geometry, or equivalently inferring that it is the total universe without the Brain material. Property: Matter must only exist if there exists a geometric from which space-time can be extracted—or Matter must only exist if there exists a substance inducing the field which induces a function of geometric instructions, proportionate to that of the geometry of the Brain material.

The degree of maximum energy is governed by the dynamics of proportioned geometry: Conservation of energy is scaled by this field of the Brain material, hence, energy is conserved by the universal force, or energy equals the unit of maximum curvature scaled and limited by the sets of the proportioned geometric. That the unification of all material manifestations, give rise to a mathematical connection of three values: space, time, and Brain substance, resulting in a framework of differentiation. The mathematical relation of the three quantities, represent the extrapolation of the universal force and of the geometric function, such that, all matter and energy shall be reduced no more fundamentally than thru the equation of proportioned geometrics. Therefore, the universe becomes a unified function, solvable by geometry and substance; substance whose origin is beyond space-time.

Of course, the new force field to which I am referring is not apparent, however, if I were able to become an observer outside the domain of awareness and perception, more accurately, to become a real hinge of space-time, I would clearly notice the powerful field of this proposed universal force. The graphical representations of the successive frames of the maximum point for geometrics to govern energy, displays the inescapable force of proportions. The field, therefore, implies the relationship of, first, the substance Brain, and second, the geometrics of the Brain material, and at bottom, proving that pure energy, or pure manifest matter, has been forced into such a geometry, more accurately that is, a proportion of the Brain material and geometry.

Space-time is a product of an interaction, and then reaction of the Brain substance and of the pure state vacuum. Super string theory contains elements which innovate a partial solution to the unification of the universe, however, to extract the most fundamental framework, beyond probabilistic and general mechanics, the final equation must contain the quantity, geometry, and presence of the Brain material. I am not satisfied with this model as a final explanation of reality, and for reasons above all that seem to be the core of unification. Derive an equation that exhibits space-time origin; that, at the absolute elementary level, one can observe the connection of that geometry and of the Brain substance.

In short, an equation that is composed of an open, changing variable which therefore permits the differentiation of matter. To illustrate, if I were to stand on Earth, or shrink down to the quantum level, or grow out into the cosmos, the force field described would be constant; space-time would still exist by proportioned geometrics, and the force field would not vary by any infinitesimal degree.

Property: The Brain material is not part of the space-time fabric, it is an illusion of the limits of idea and perception. The illustration displays the limited imagery of the placement of the Brain geometrically in relation to the space-time manifold. The Brain material does not accelerate, nor occupy any space points in the universe, therefore, the unifying field equation predicts that velocity, change, distance, and thermodynamics, are properties of proportions, and not of the regions behind the surface of the Brain material substance.

Upon the Brain material interacting with the vacuum state and extracting the fabric of space-time, the expansion is nearly exponential, however, here is the fascinating part; that the bound of such a geometry is not infinite nor finite, more precisely it is a unit in between the two limits, occurring as a result of the dynamics of proportioned geometrics. What is in question, from observation of our perception of nature, in that the Brain material and matter appear to exist within the equivalent structure of space-time, is how and from what process does such a grand illusion manifest? Within the limits of idea and perception, such false impressions develop perhaps as necessities, nevertheless, the objective at this time concerns the geometric order of the Brain material and that of its effects on the physical universe.

Proportioned Geometrics & the Brain material
Part II

What the unifying field equation holds to be true, is that, the universe, in order to conclude, must be broken down to pure geometry, in that, no further reducible functions be derived. Second, that the nature of the substance in conjunction with geometrics, contains origins of such enormous power and energy conditions, that its own materialization deifies all preconceived limits comprehensible to thought. This substance, as I theorize, has become broken off of a collective unit prior to interaction with the vacuum material.

The Brain material as depicted graphically, occupies a bound which that of the extracted substance enters, and hence, a most intense interaction occurs. The unit, Brain substance, and that of which the name vacuum may be assigned, a region yielding zero space-time, zero differentiation, zero energy, and zero geometry most importantly. The geometry—field induced is such, and thus if such, then there is an equation, for all Brain quantities, which sets the described space-time fabric, for all physical scales, for all times, and for all empirical observations. Thus, from this proposed equation, formula, and expression of geometric variables, a new force is derived, a universal force defined as the result of the presence of the Brain material.

Geometrically, a point can not be mapped from outside space-time, and reasons being, that the Brain substance, under proportioned dynamics, would clearly make this unlikely. Suppose a system of particles being smashed into one another at extremely high velocities, and are reduced into much smaller constituents, and similarly, if a cosmic bat were to interact with a cosmic object such as a star, and are reduced into smaller constituents then conservation of energy laws may be proven under the equation of unified fields, and such that there is a condition for any system in space-time to be observed that constrains its geometrical and mathematical elements to be not an independent function, rather, of a proportioned scale governed by the proposed inescapable field of force.

This solution to the unification of the universe is such that, to materialize to a structure of space-time, there must be dynamical geometrical proportionalization of fields; there must be a quantity separate from the laws of physics, one that can contain immeasurable units of force. The order of this solution in relation to the principles of dimension: In observing space-time, I may question the limits as to how large or how small matter can be defined, geometrically, thus measuring such would produce results not consistent with uniformity, instead, the results would yield a framework missing a most crucial centerpiece; what yields is a geometry with seemingly independent variables at different distances and times. The goal is to comprehend our physicality not by deriving separate laws for separate times and spaces, the goal is hitherto discover the function of all matter and constituents, and derive a solution which must reduce and unify under the most absolute fundamental geometry.

The corresponding illustration of the field induced by the presence of the Brain, matter, light, and the unifying equation, are collectively depicted as the sewing of the ends of space-time fabric, by the dynamics of proportioned geometrics. Observation: As I seek further into the framework of the origin of all such regions of nature, the connection between the boundary of the Brain material and the maximum points of the surface of the interacting vacuum, that if space is therefore proportioned, and as declared, that the Brain material derives the function of differentiation, then the property of time must as well be proportionate to the geometrics of the material Brain.

Maximum limit is a variable in the unifying equation. Thru the dynamics of proportioned geometrics, the field in which the total degree of unit tension on the surfaces of matter, induces material multiplication; an electron, now interacting under the electro-magnetic force field, contains the force behind the force, the universal force field caused by the presence of the Brain material—governed by the laws of proportionality. Energy interacts due to one unvarying force, the extraction of space-time integrates energy into a function of geometric proportions, thus there are no methods of suppressing this universal force field. The particles of light can not escape the force at its velocity, and now from the principles from the geometric framework of the Brain material placement, which does not occupy space-time, the speed of light would not be constant, but varying.

The universe is a paradoxical unit, and in order to solve its origins, an equation connecting the equivalence between the Brain material geometry and the units of space-time must be incorporated as a standard. Sealing the ends of space-time for example, are marked by the curvature caused by the presence of the Brain material, this sealing is a direct proportion as displayed in the corresponding illustration.

The expansion of space-time thrusts outward per the interaction of the Brain substance and the vacuum state, and therefore, the actual radius of the universe is proportional to the radius of the Brain material's geometric radius, however, the geometrical expansion of space-time is finite, in that, the rate at which it expands depends upon the rate at which the Brain structure expands, except the rate at which it is expanding at the present time, would not slow down within the range of an average life time.

If an insect were to observe the geometric properties of space-time, they would derive the equation of unvarying unification, or if a larger cosmic being were to observe such properties, they would derive the identical equation, except that one quantity that would differ from observer to observer, would be the rate at which space-time expands. The geometry would be perfectly proportional, but the rates would differ. The change in rates from being to being do not affect the equivalency of the principles derived from proportioned geometrics; the change in rates of expansion only show how light when entered in one observers equation, may take slightly longer or shorter to reach a point in space-time.

Universal force applies homogenously, uniformly, and isotropically, and that such is inescapable because Brian substance interaction with the vacuum state, hence any force defined in nature, is broken down to the geometrics of proportioned dynamics, and that the probing of constituents further than those of the geometrics scaled by the Brain material, leads only to the yielding of the unifying equation at all scales. It is necessary to understand the concept that the Brain is a substance, which to our senses, is not an obvious structure that is ultra energized beyond any known measurement; it is a system which contains properties of such magnitude to yield perception. The Brain material can be viewed as a non-biological system, and can be defined as a super substance with direct geometric influences on the physical universe.

Proportioned Geometrics
Part III

If I were to see the universe with detection goggles capable of measuring the super density of the Brain material, then it would be clear that such an object from all distances equivalently radiates a massive glowing field. This effect is most likely caused by the intense charge of the material, and by the actual aftermath of transformation from the still unknown collective manifold, from which it broke off, and then of the interaction with the vacuum and pre-matter states. The shape of the universe, as a whole quantity, contains a formulation dependent upon the dynamics of the Brain material shifting into a state of direct interaction with the vacuum; thus at that very early instance, space-time undergoes a rapid expansion, and from this moment of geometric manifestation, the degree of curvature is defined as proportioned to the curvature of the Brain topology.

If the unifying effect can never be escaped, then at all instants within any observed system, the field equation would produce the identical force. This relation and force, if I were to ask for a reference as to quantify the strength of it, quite simply, the measurement results in perception. The material is so incomparably charged, that the expansion bursts out into the form of a perceptive quantity. One central concept, in order to correctly piece this theory, is the dynamics of extraction, that the geometric proportions of energy are only determinable by the precise field limits that the geometric Brain material sets.

Matter must originate as a composition of a sort of non-geometric fabric pressed up against a vacuum by a parallel duality. There exists an internal component within the equation, that which would define the coordinates of pre-matter. The order of such pre-matter units transform instantly to a geometrical set once the Brain material interacts with the vacuum and pre-matter states, that is matter. To obtain the transform of the quantity pre-matter to the set of matter fields, the function connecting the two manifolds, that of the vacuum and that of the pre-matter sheet, needs to be defined thru proportioned geometries; the Brain material provides the required substance and geometrical dynamics to initiate space-time-matter transforms.

The Principles of Perception & The Initial Conditions

George Berkeley, the great thinker of the eighteenth century, proposed that matter is nothing more than an intangible idea, which therefore can not exist but only if there exists a mind to perceive it. This theory is correct, in that, it defines the relationship between the mind and the material world; it opens the margin for the limits of how and to what extent an idea may be stretched in our capacity to understand. Generally, for instance, given an external system, and if considering the impression it makes on our senses, such that, if a flame is hot the idea and sensation on the mind is that of temperature. If I were to discover an object which reflects light quantities, then the idea derived by the mind is that of light brightness.

The question which results from Berkeley's theory of perception, is that, if the material world generates only ideas about its structure, then what is matter when the idea of it is therefore stretched beyond the limits of impressions? More clearly, if I measure heat, I am not determining the nature of such a material quantity, in its place the measurement only further proves the idea of heat and not the core of its placement. Another example is a fundamental law: that all energy is conserved, such that, thru any interaction, matter can not exit the fabric of space-time. Thru the same methods of reasoning, it becomes apparent that any and all attempts at deducing information about the physical universe has to be subject to the limits of the constant uncertainty which the mind causes, and from which, the idea of the properties of matter arise, and not that of the absolute properties of matter within itself. The extent of the understanding of any physical system, that which is perceived, is governed by the inability to separate the idea of matter from the matter that is absolute in form.

The initial conditions of the beginning of the universe, from the current models, are those of extremely hot super dense interactions, further cooling and forming the expanding system observed today. There exists a limit, such that, for any system within space-time, and that the proportioned geometrics have been defined, increasing any variable of a material system, that of temperature, velocity, mass, area, distance, direction, can not transform the unifying field equation, which contains the components: Brain material, vacuum (nothingness—zero space-time—geometry, etc.), the terms of the interaction, the functional transform of pre-matter to matter geometrically, the universal force field components, and the sets of varying rates of expansion, and the total shape of space-time set proportionate to the geometric order of the Brain substance.

With respect to the initial condition requirements to extract space-time and derive an underlying equation determining the total curvature and geometrical value of the universe, the enormous value of temperature at the point of induction of space-time, as identified in the Standard Model, would not be sufficient strength for that type of material production. The only material containing the values of power to induce such conditions, are that of the Brain structure interacting with the vacuum and pre-matter fabric. It is at this point of interaction, that the fabric of space-time undergoes geometrical expansion, and at which the shape of space-time is fixed according to the Brain material configuration. There exists a sphere of influence, occurring simultaneously, at and from which the instructions for material transforms from pre-matter to matter geometry transpire.

The contour of space-time is therefore dependent upon that of the locality of the Brain material; there is only one way in which the Brain material unit may proportionatly extract the total curvature of space-time, and space-time itself. If to understand the geometry of such, consider a sphere, first, convert it into a torus (donut shape, etc.), the Brain substance inserts into the middle, and space-time expands to a massive size, such that to observe today the shape, it only appears flat, when conversely, the Brain material curves the universe. It is as if inorder to view the true nature of space-time origin, biology must be completely unobserved and disregarded.

The Brain material contains geometric, mathematical, and substance properties, which together, construct a framework under which the extraction of space-time and of the transform of matter from pre-matter, connect the influences of the fundamental forces to one underlying force field: the product of the Brain material interaction with the vacuum and pre-matter states. The root of the unifying field equation is from that of the exact instant of interaction of the Brain material with the two elemental states above, and at which, the interaction of matter may occur due to the transform. Matter without geometry is nothing.

The Components of the Unifying Field equation Part IV

The pre-matter quantity does not contain space-time coordinates, nor contain geometry, and it is one of two parallel units, the other is that of the vacuum state. These components induce a parallel relation because they have nowhere else to go, and at the instant where the Brain material shifts from its pre-manifold quantity to interact with the pre-matter and vacuum states, that the universe functionalizes, and the transform occurs of pre-matter to materialization governed by proportioned geometrics, consequently, the given components unify under the mechanics of their geometric placements; the above quantities continue as unbreakable parts thru such a connection. Following the interaction of components, differentiation and integration are then capable to occur as a universe divisible and transformable under the equation of proportioned dynamics.

The nature of the origin of the Brain material is that of a genus apart from the generateable idea, furthermore, the idea would have to be stretched beyond our capacity to perceive or imagine, and the collective pre-Brain substance occupies a central domain exterior to that of the dynamics of an idea. George Berkeley contested that, "to be is to be perceived", that the very essence of something is in that of it's capacity to be perceivable. At bottom, it is that of faith which will lead this inquiry to a superior conclusion, and that which takes precedence, is that of the unseen workings of God.

The following pages are illustrations of the proposed hypothesis. Each illustration contains a number corresponding to the descripition of each as detailed below.

1. The middle semi-spherical structure is the Brain material mapped out geometrically, and displaying its nature as a quantity of geometrical order. The hyperbolic lines increasing exterior to the Brain structure, are the product of the interaction of Brain material and the vacuum state, space-time fabric and the arcs depict the transform of pre-matter into matter as governed by the universal field of force. The Brain material is not located in space-time, and consequently does not propogate or accelerate.

2. The middle semi-spherical structure is the Brain material mapped out geometrically, and the vertical wavy object crossing it, is the Brain material as it illusively appears to an observer. The co-ordinate cross sections are the extractions of the space-time dimensions.

3. This representation contains an arbitrary expression meant to be the unifying field equation, and the progressive curved pieces show that any system can be derived from this initial function. The components of the unifying field equation are: The Brain material the geometry of the Brain material, vacuum state, pre-matter fabric, and instant of interaction variable of all three.

4. The Brain material's geometry.

5. The curved line influenced by the changes of the tanget lines, represent the universal force determining the materialization of matter geometry.

6. The wavy object to the far left is the Brain substance previous to entering and interacting with the vacuum. The two converging lines show the grand condition of the very point where the Brain interacts with the vacuum state, and parallel to configuration of pre-matter.

7. The random squiggly lines denote the idea of an idea contained by the limits under which it can not escape. The individual line at a point, is the region in which an idea would have to occupy to increase our conception of what pre-matter and pre-brain material are, fundamentally.

8. The entrance of the geometric Brain.

9. Two comparative charts containing the four fundemental forces, gravity, the strong and weak, and electro-magnetism. The left hand side is the correct order of combination under one unified field equation, and the right hand side is the incorrect method.

10. The Brain duality as a geometric composition, and next to it is the powerful super charged substance.

11. The geometric Brain located in this tunnel, originating from when it first interacts with the vacuum and pre-matter, and it is the true and exact distance in the universe.

12. The transform of pre-matter to matter.

13. The vectors to points demonstrate the invisible surface of the topologies from the Brain existing.

14. The extraction of the number and value system, as defined by the presence of the Brain geometry.

15. The Brain and the surrounding field it creates; the structures are the energies in space-time being forced to shape as a proportion to the geometry of the Brain.

16. The illustration of the Brain material inhabiting a region behind space-time.

17. The constant force field induced by the Brain, and the energies, as represented as particles, compounded by this force.

18. The concept of cutting out the edges of the Brain, geometrically, and observing the impact of such on the rest of the physical universe.

19. The intersecting spheres represent the mathematical significance if the sum of all mass in the universe be compacted, that the rate of space-time expan-

sion would not be affected as a result of the unifing equation and the Brain location.

20. The Brain structure, and adjacent are the proportioned outlines of energy and matter.

21. The affect of the power and super density of the Brain material, such that, the material glows or radiates a frequency beyond any measurable regard.

22. The multi-wavy connecting objects are the Brain material up against the divide of vacuum and pre-matter topologies.

23. Frames of pre-matter material (middle), left is the geometrical matter influenced by the univeral force field, and the Brain material in relation to space-time.

24. The sequence of the pre-Brain substance interacting: space-time extraction.

25. The Brain material traversing to the expansive frames of space-time.

26. Depiction of the Brain quantity sealing the ends of the fabric of space-time.

27. The universal force as the "force" behind the force, that which is caused by the Brain occupation.

28. The Brain material, the pre-matter structure, and the vacuum state, as a sum, and causing universal differentiation.

29. An individual piece of Brain material, that which is not reducible by means of power value.

30. The equivalence of energy fields at varying scales, and the effects of the universal force field.

31. The resembling human figure and the geometric Brain in relation to the "idea" and the material universe; the illustration formulizes George Berkely's theory of Perception.

32. Objects in space-time have a total distance of zero between them.

33. The radius of the universe, as set proportionate to that of the geometric Brain material.

34. The pre-Brain manifold transforming geometrically; the radius of space-time as shown.

35. Proportioned geometrics. Individual frames of matter propogating thru space-time, and the Brain material inducing the lines of force of the unifying force.

36. The Brain material with a vector traversing it, and the hyperbolic lines diverging signify the product of the interaction, space-time.

37. The Conservation of Energy law, as connected to the dynamics of Proportioned Geometrics. Collison of particles, the production of reformed matter influenced by the universal field of force.

38. A frame of the radius of the Brain material and that of space-time.

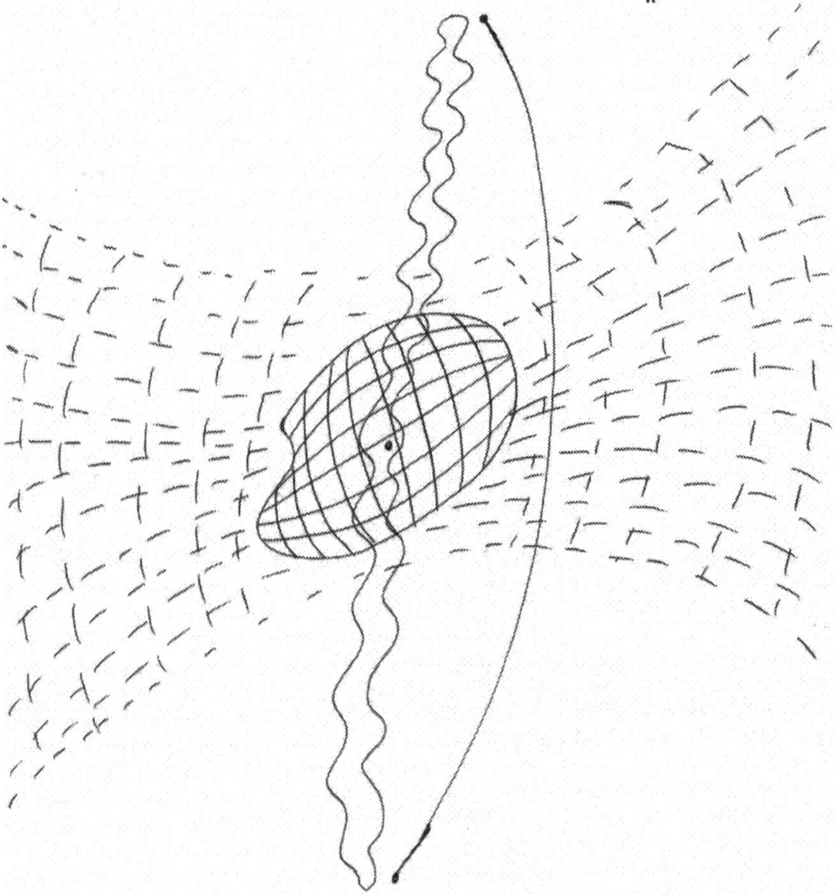

2.

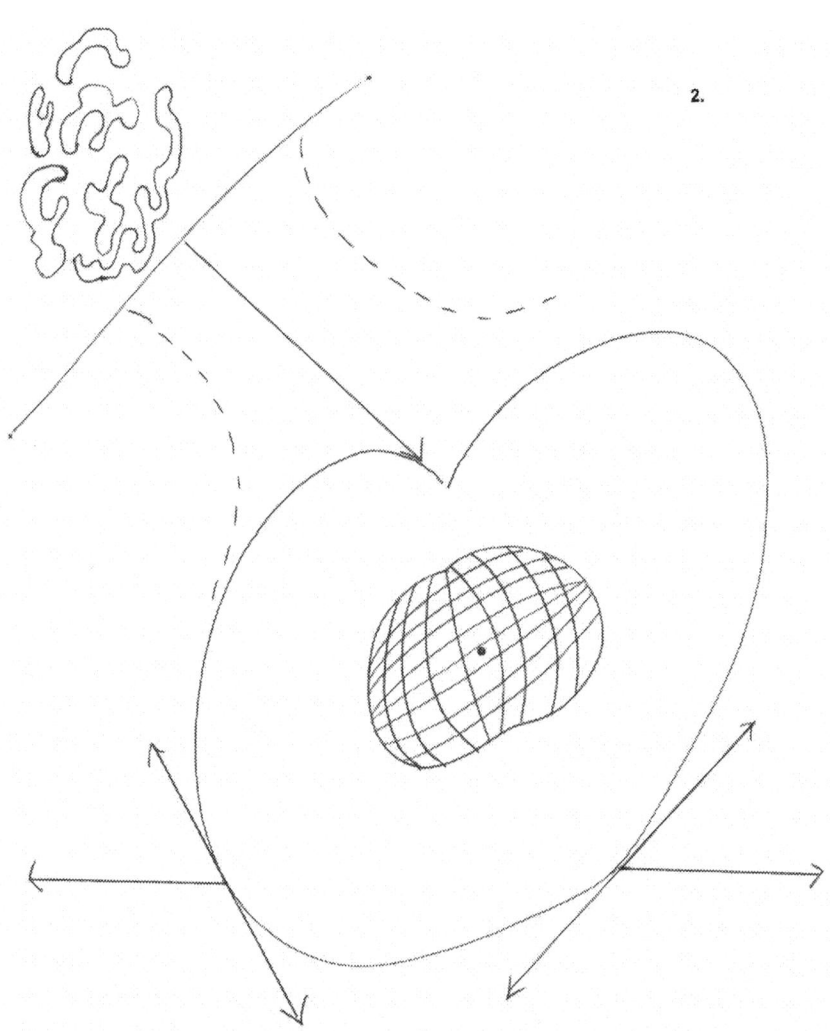

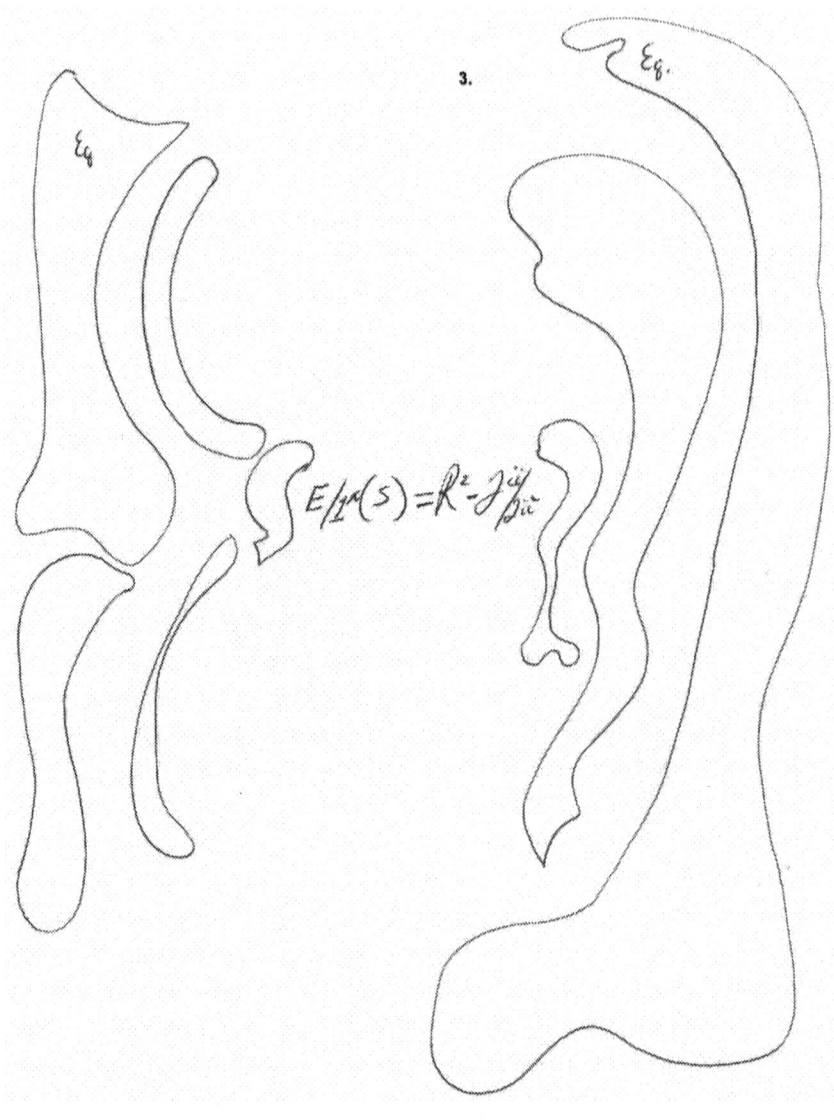

4.

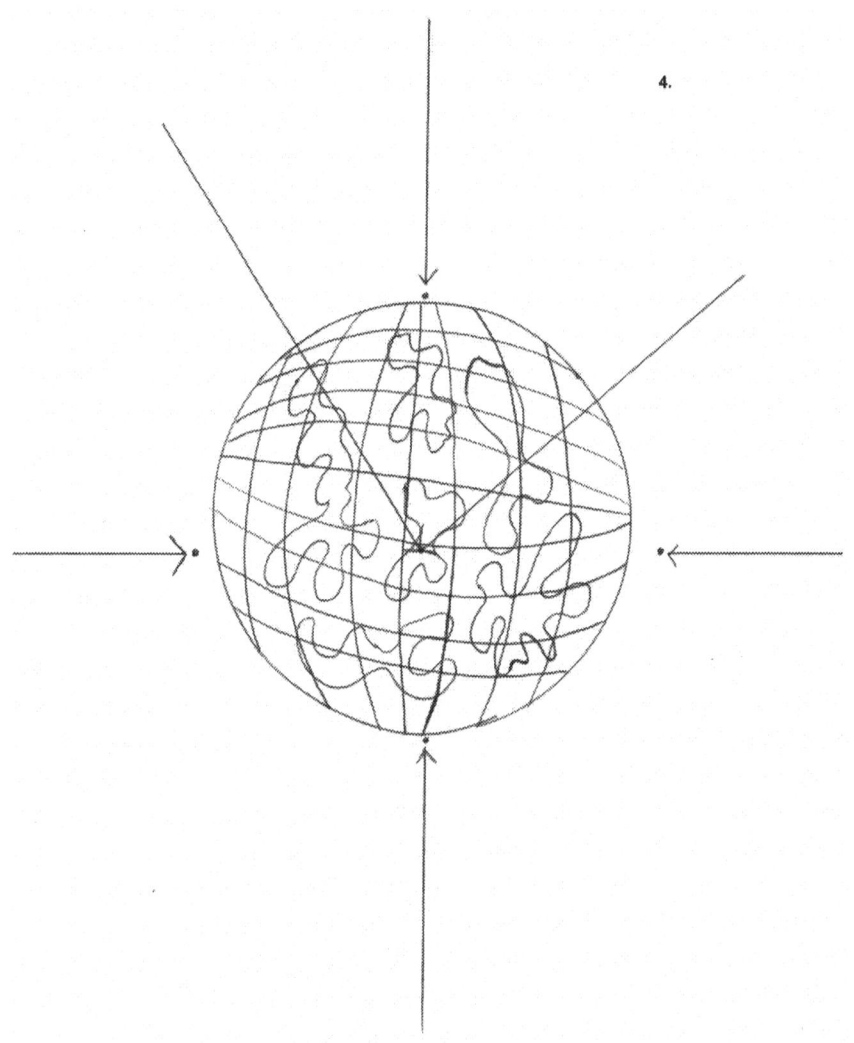

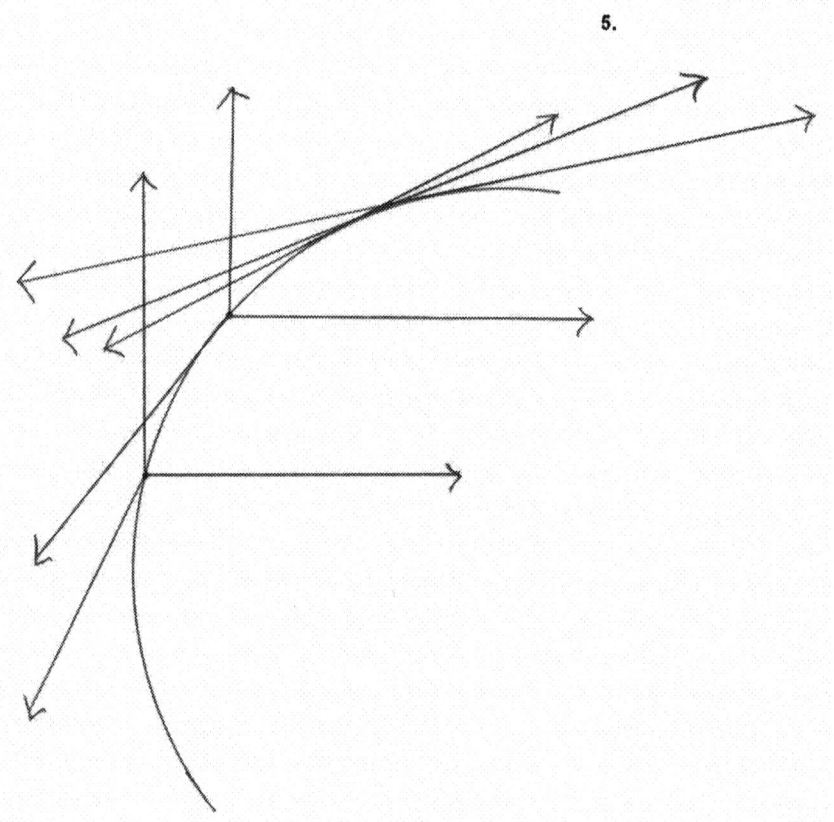

5.

6.

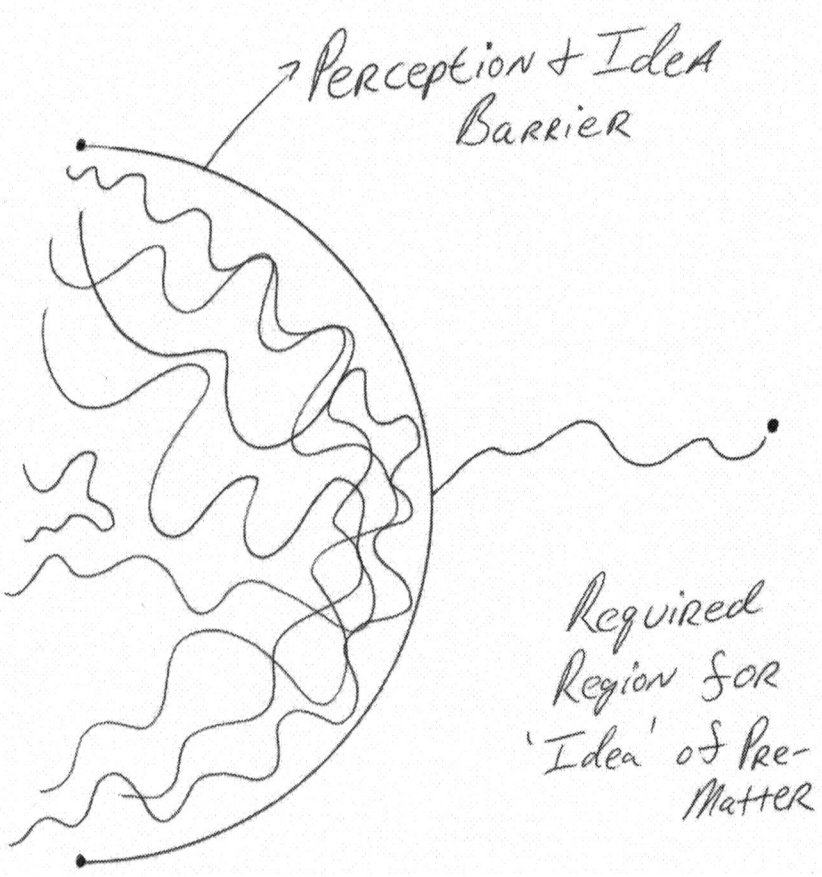

8.

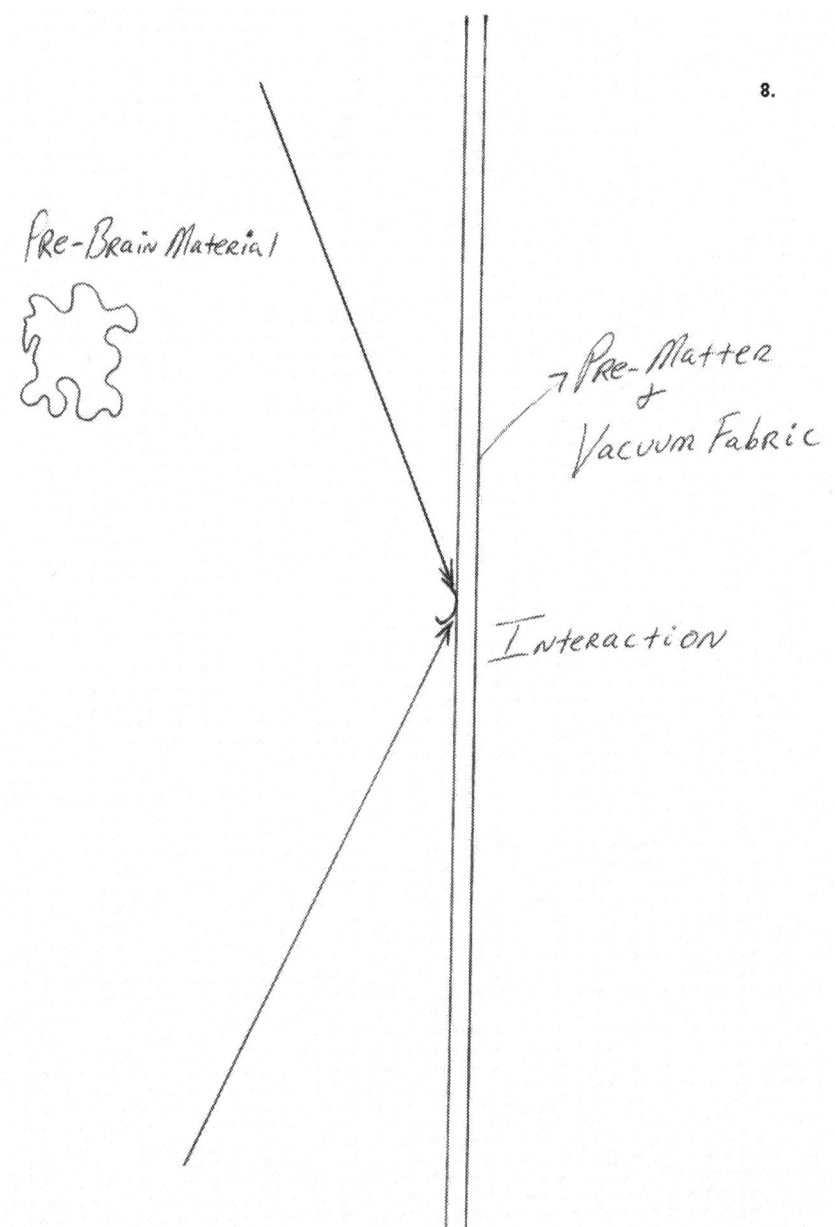

Pre-Brain Material

→ Pre-Matter
+
Vacuum Fabric

Interaction

9.

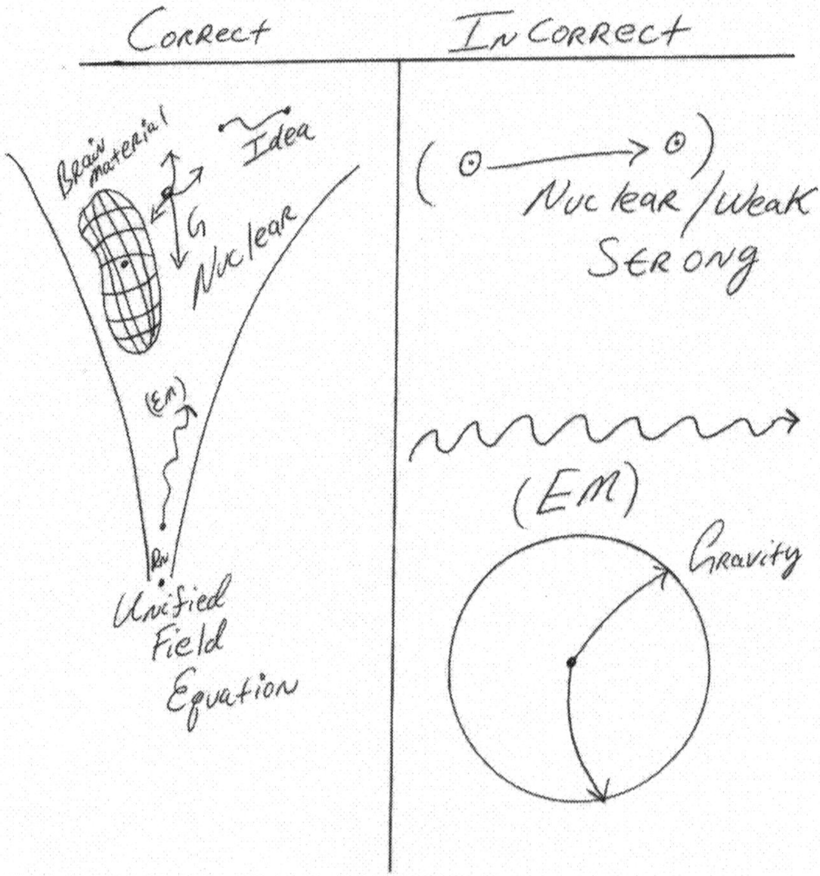

Correct

Brain material   Idea   Nuclear   Nuclear

(EM)

Unified Field Equation

Incorrect

(⊙ ———→ ⊙)
Nuclear / Weak
Strong

(EM)

Gravity

10.

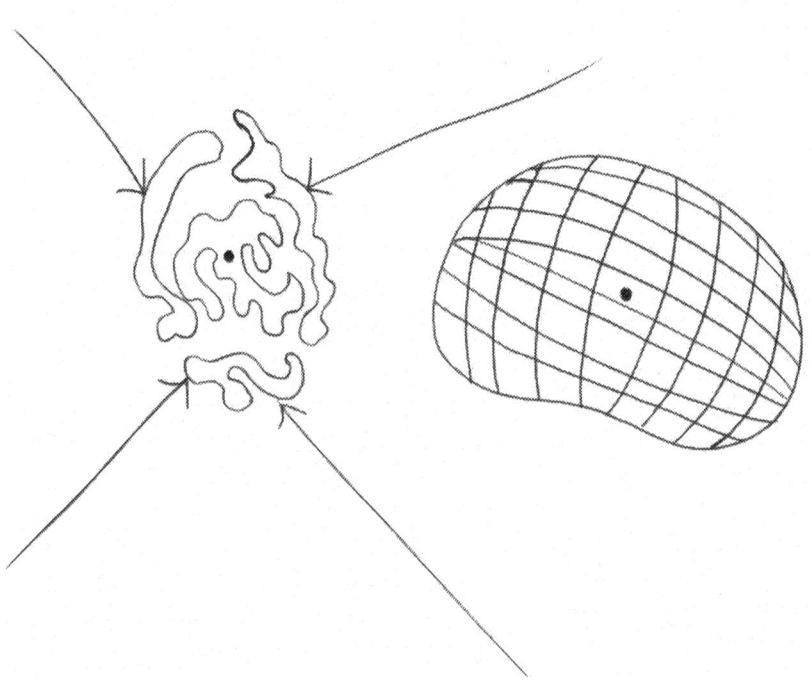

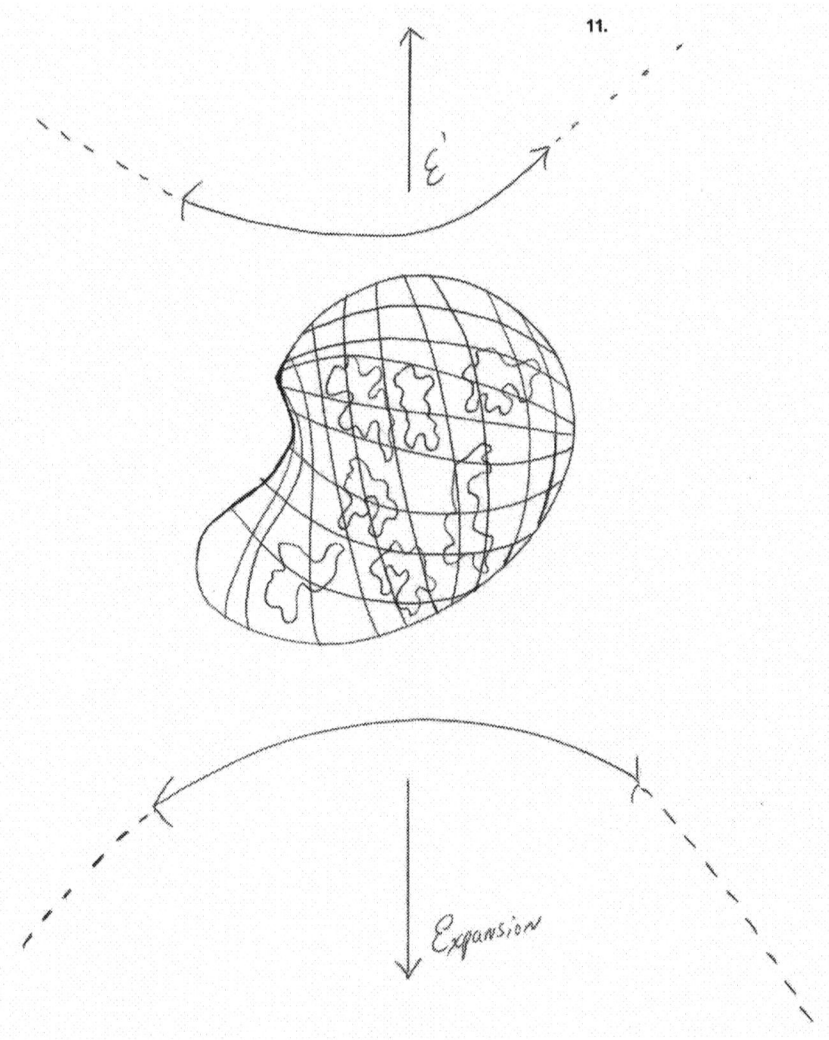

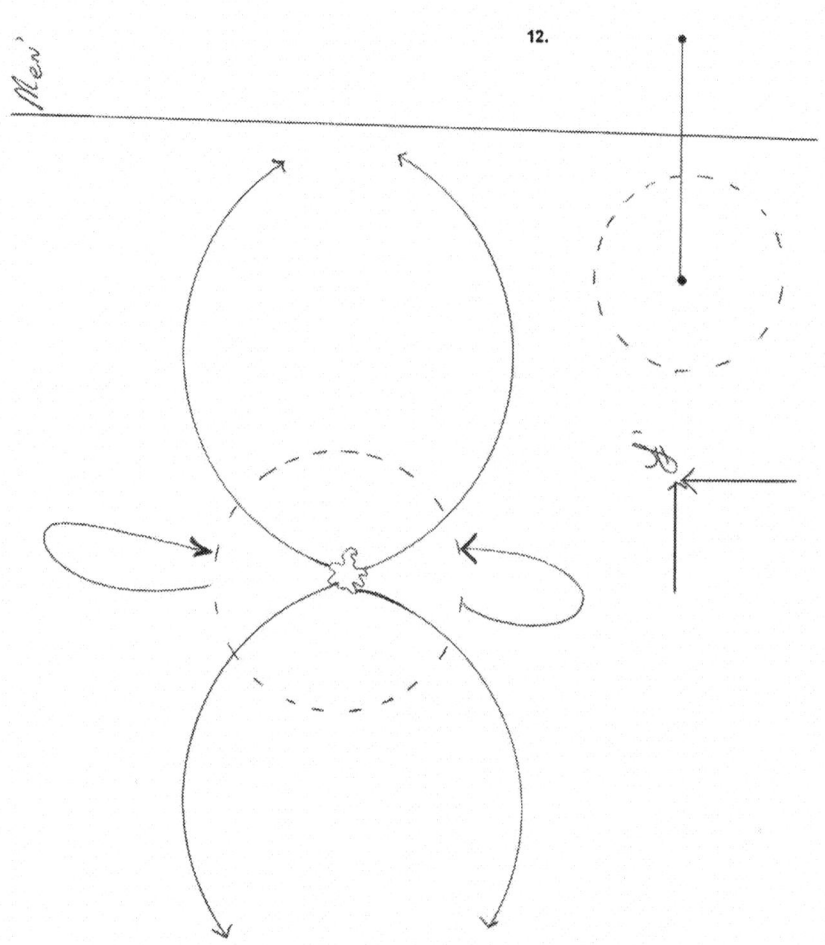

13.

Universal Force

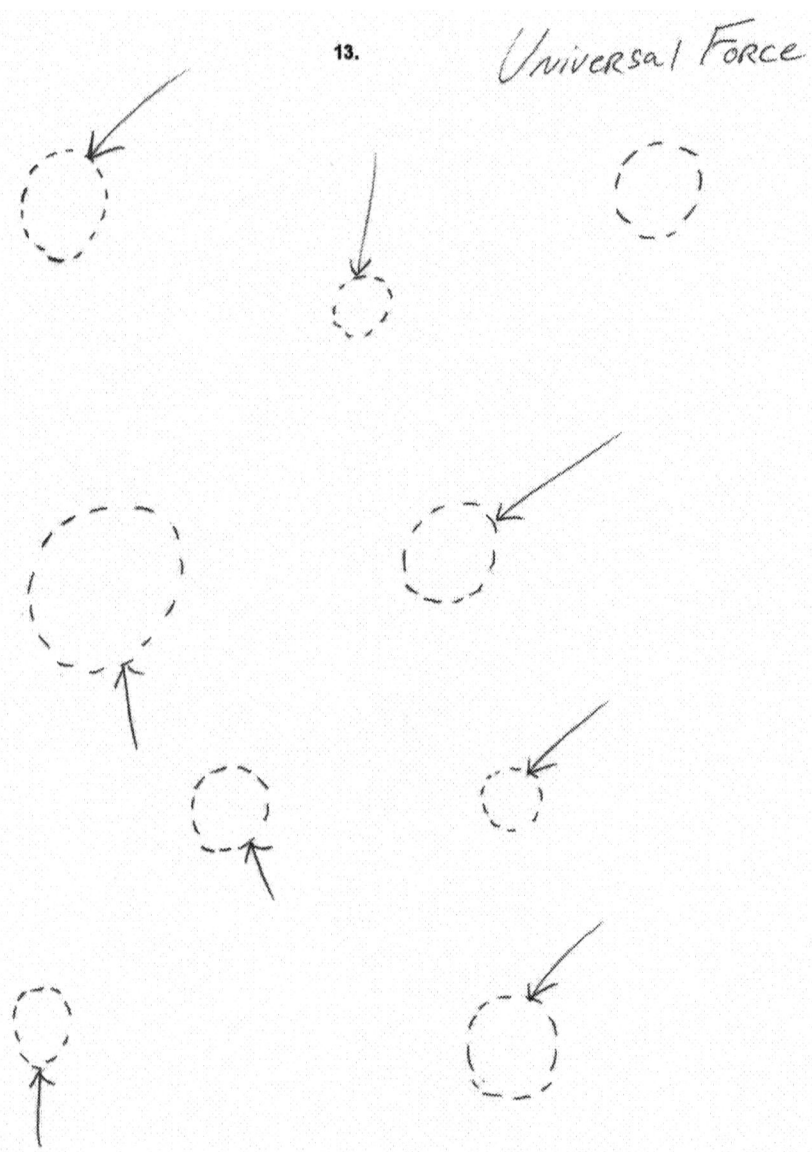

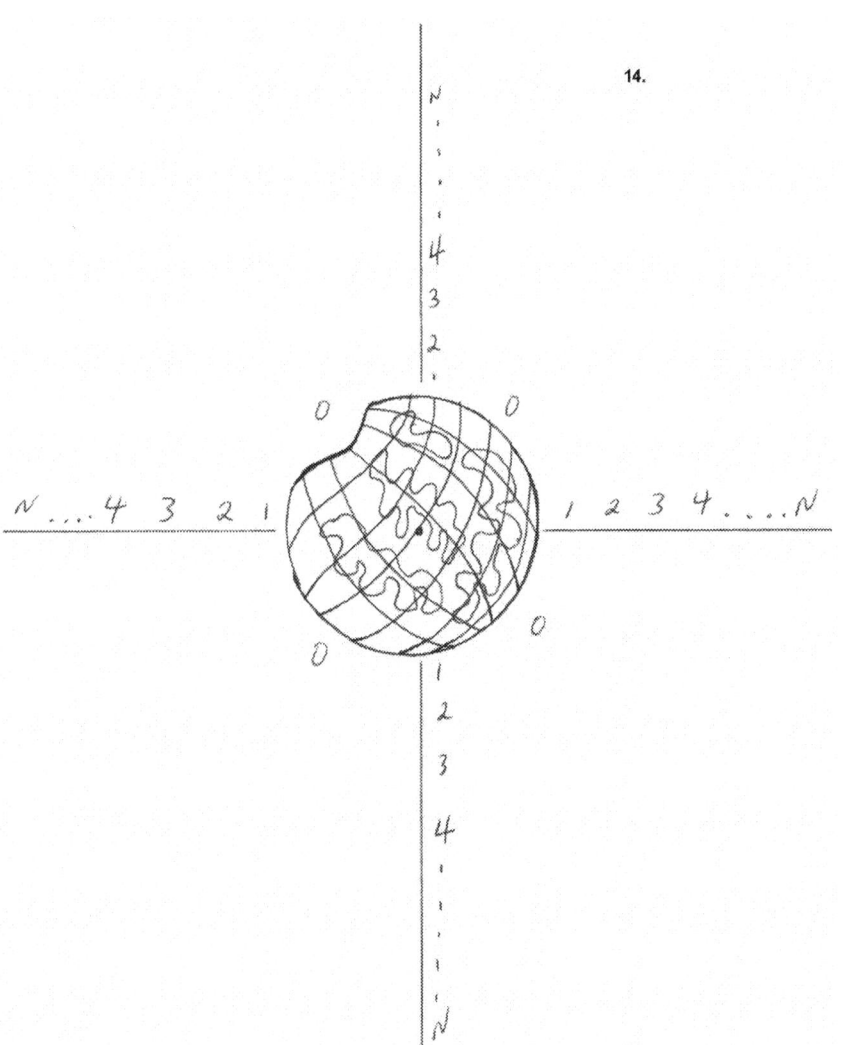

15.

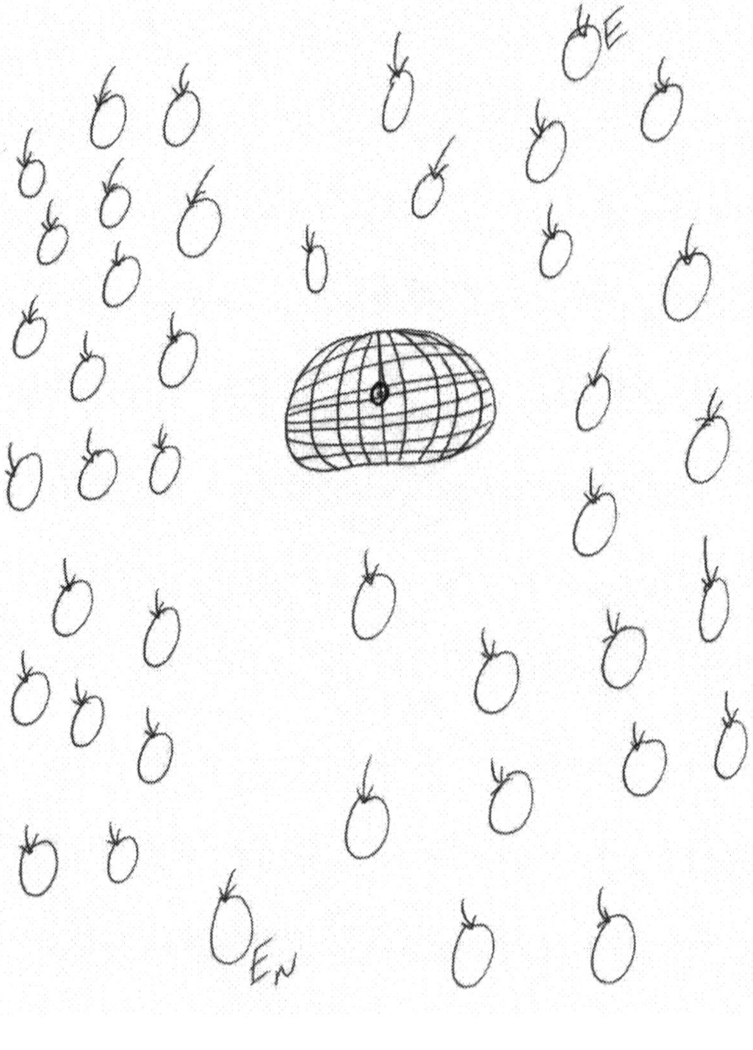

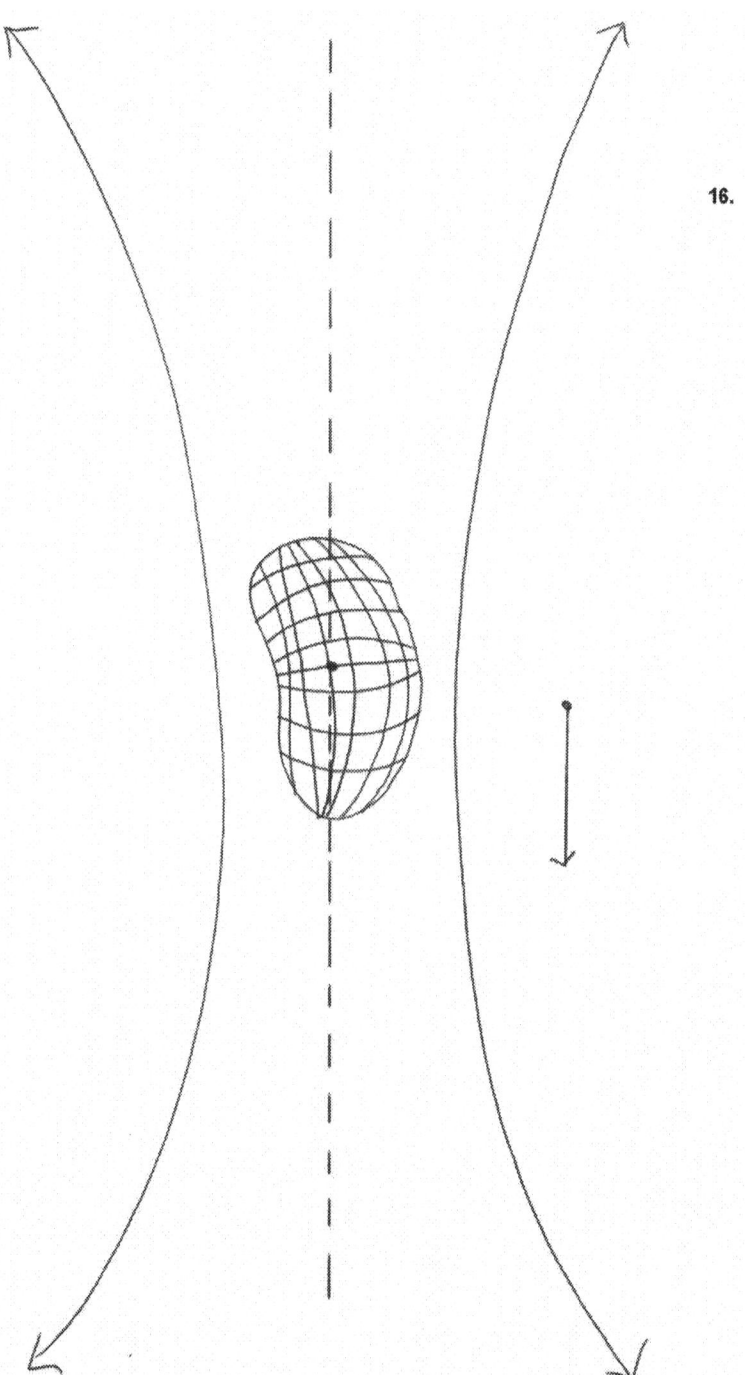

17.

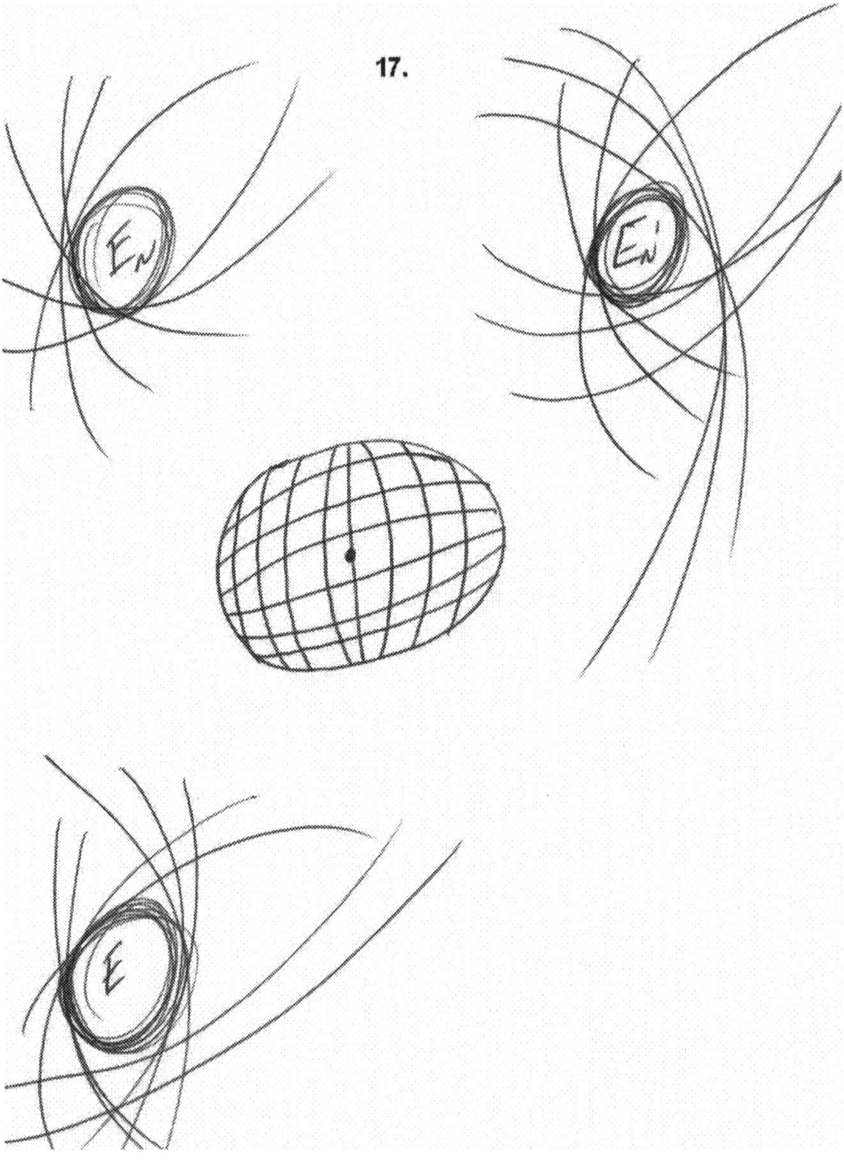

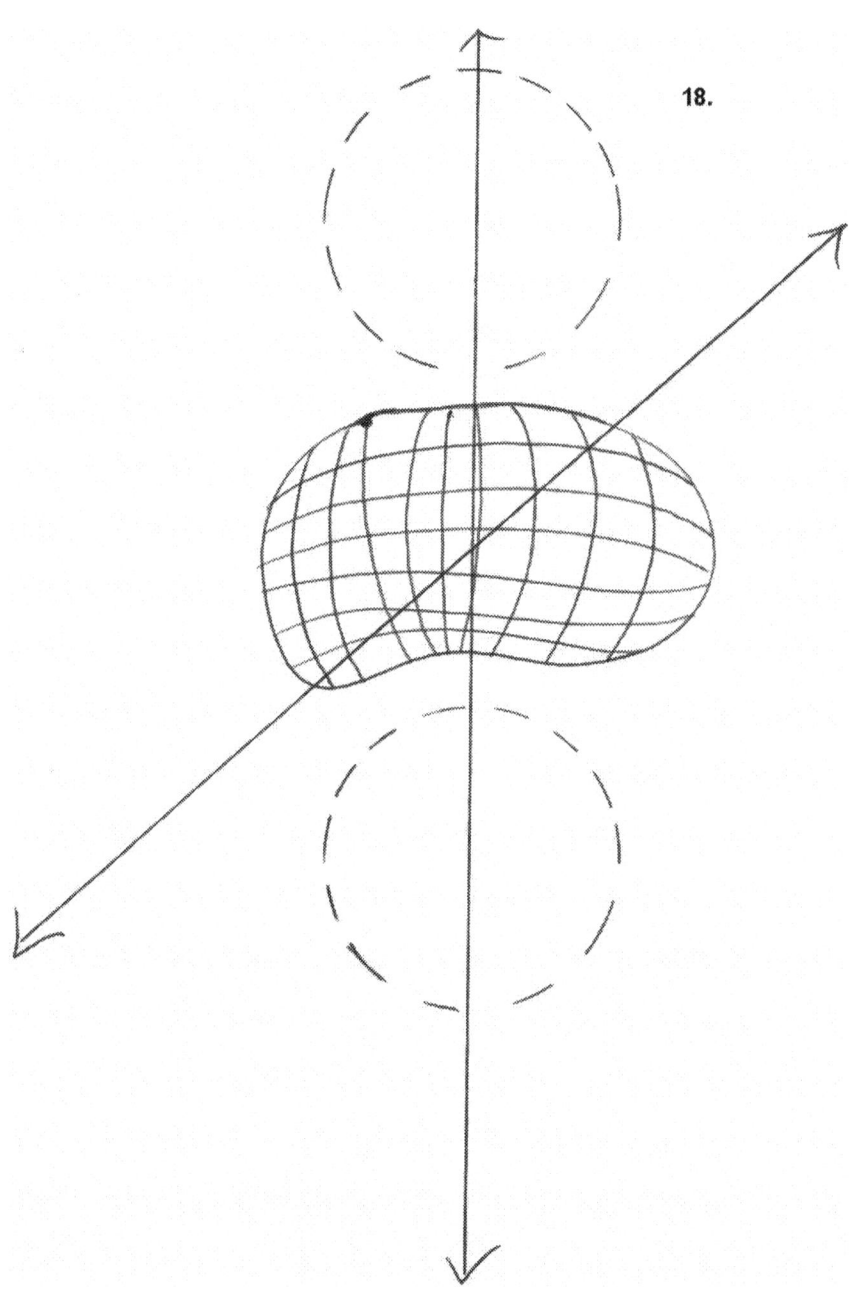

18.

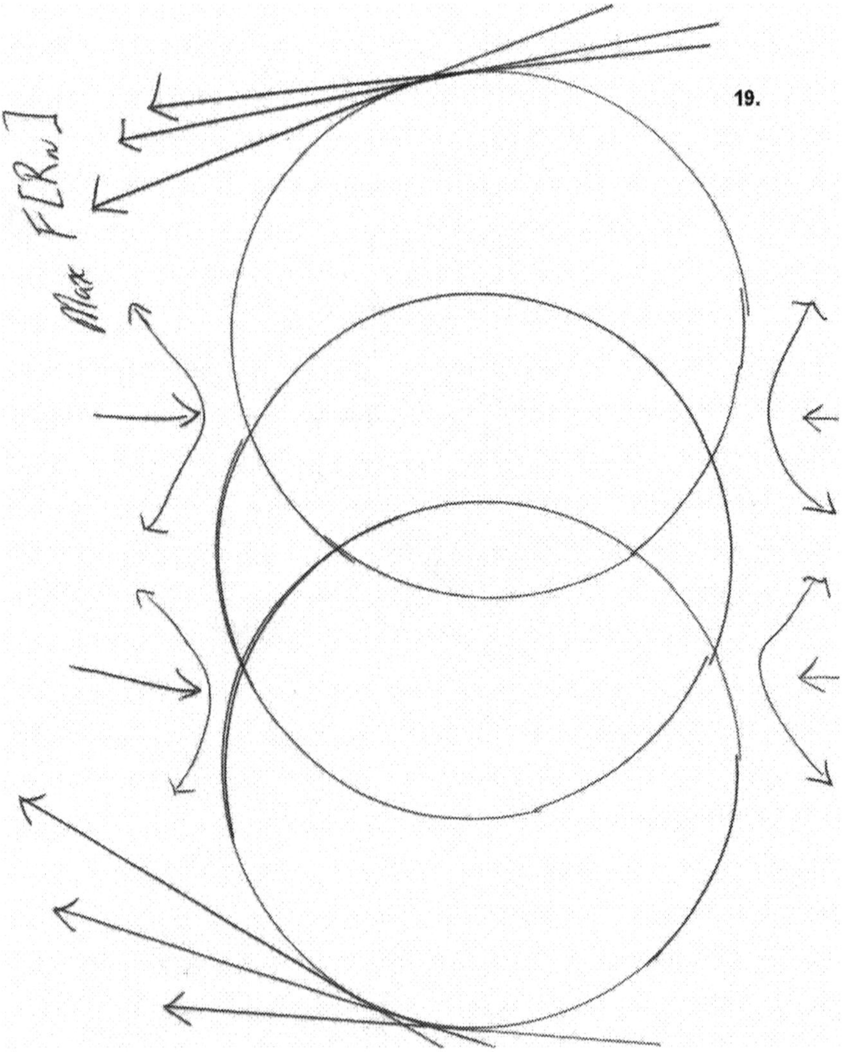

Max FLRw]

19.

20.

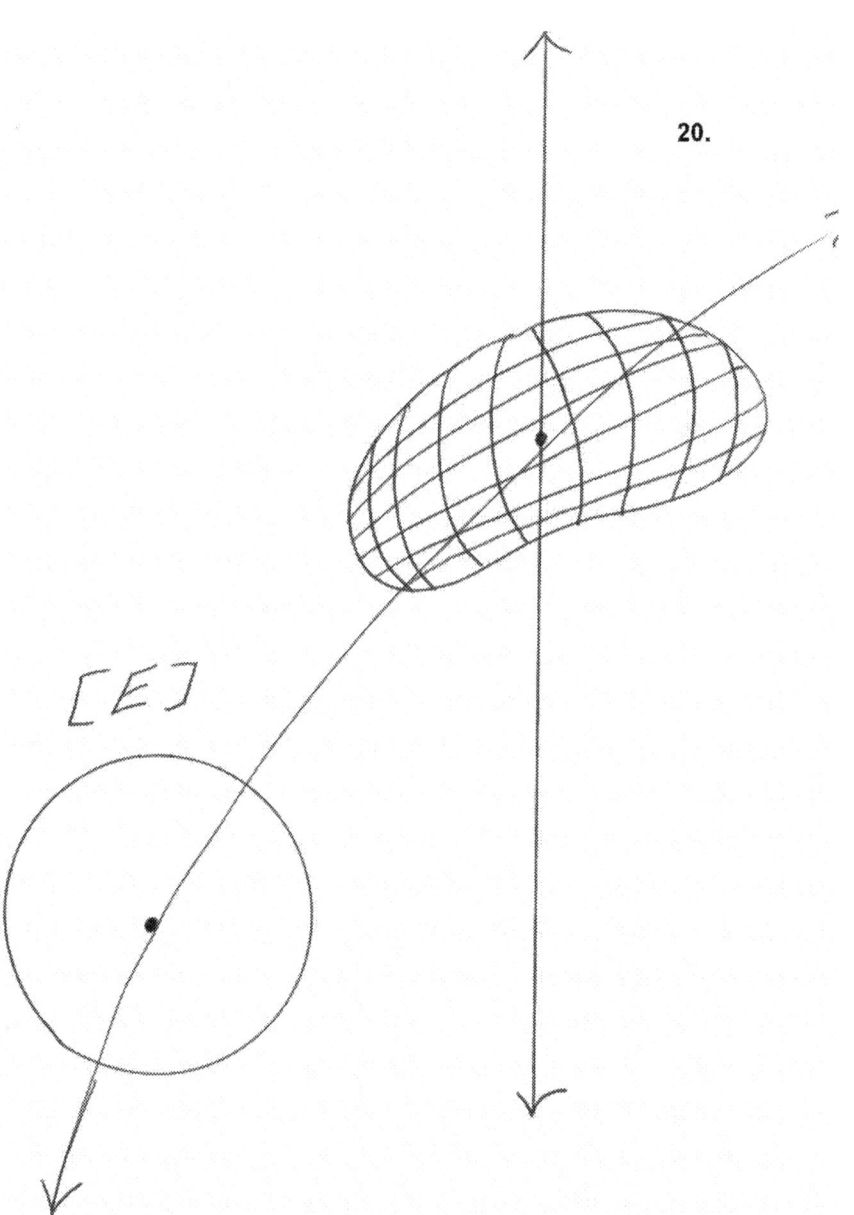

[E]

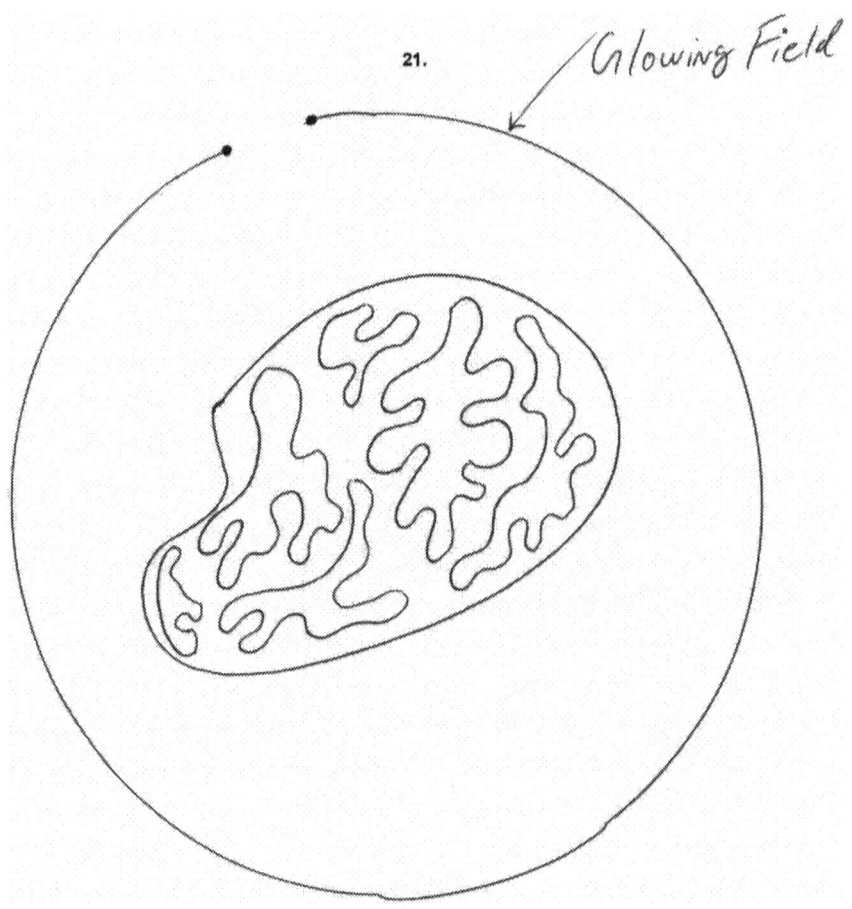

21.

Glowing Field

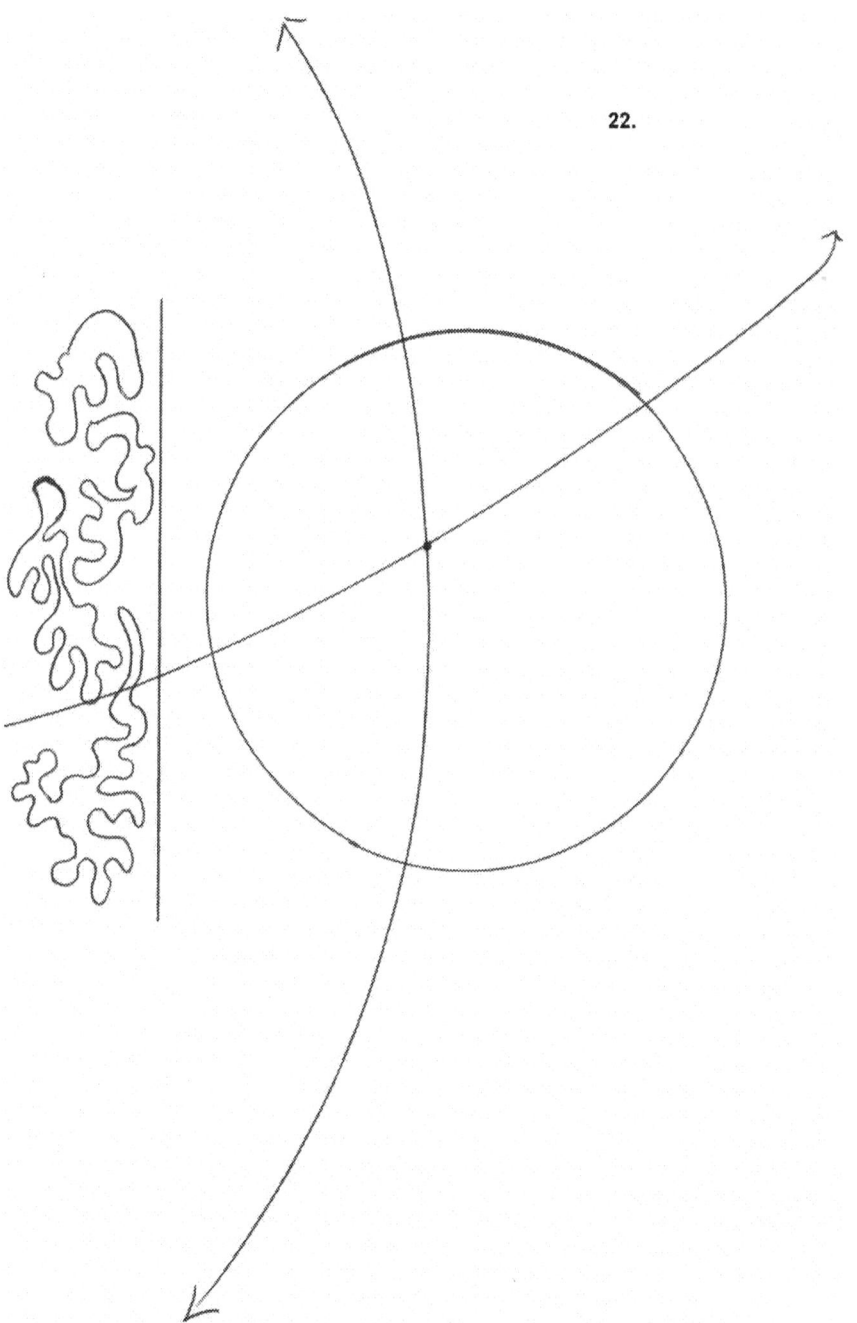

22.

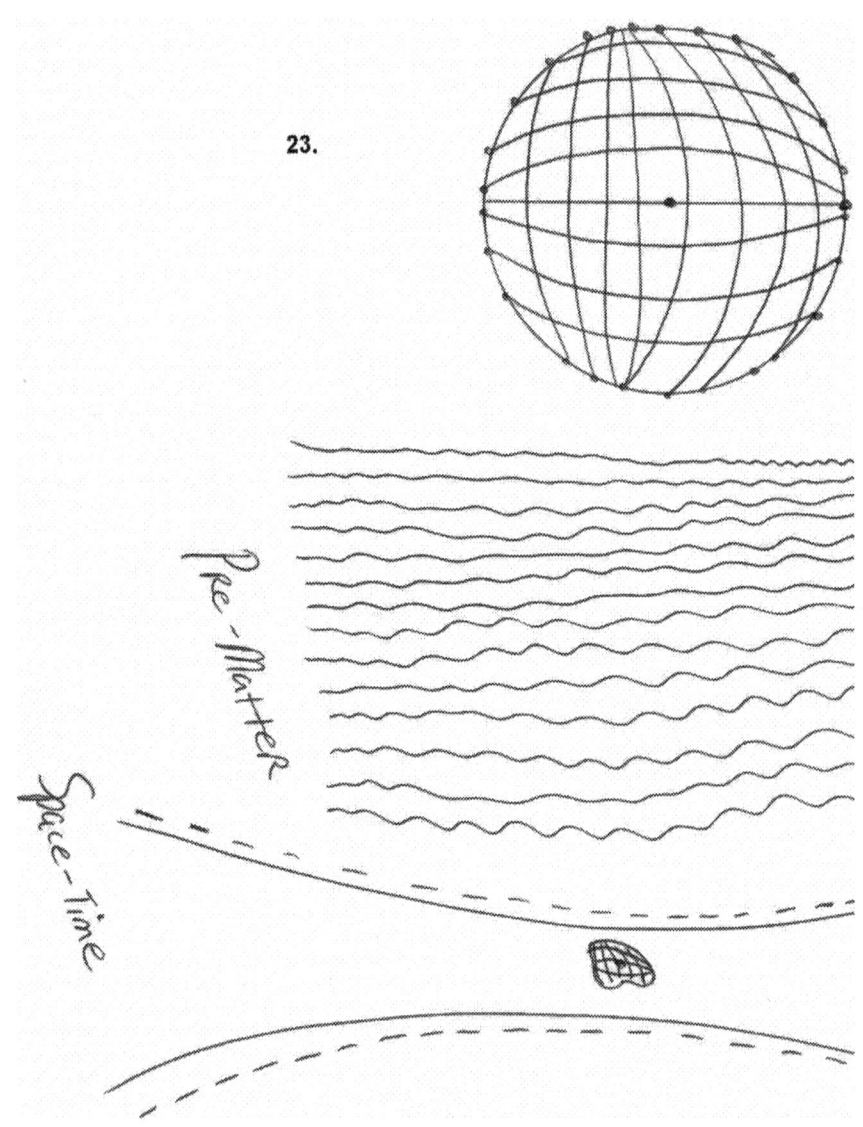

23.

Pre-Matter

Space-time

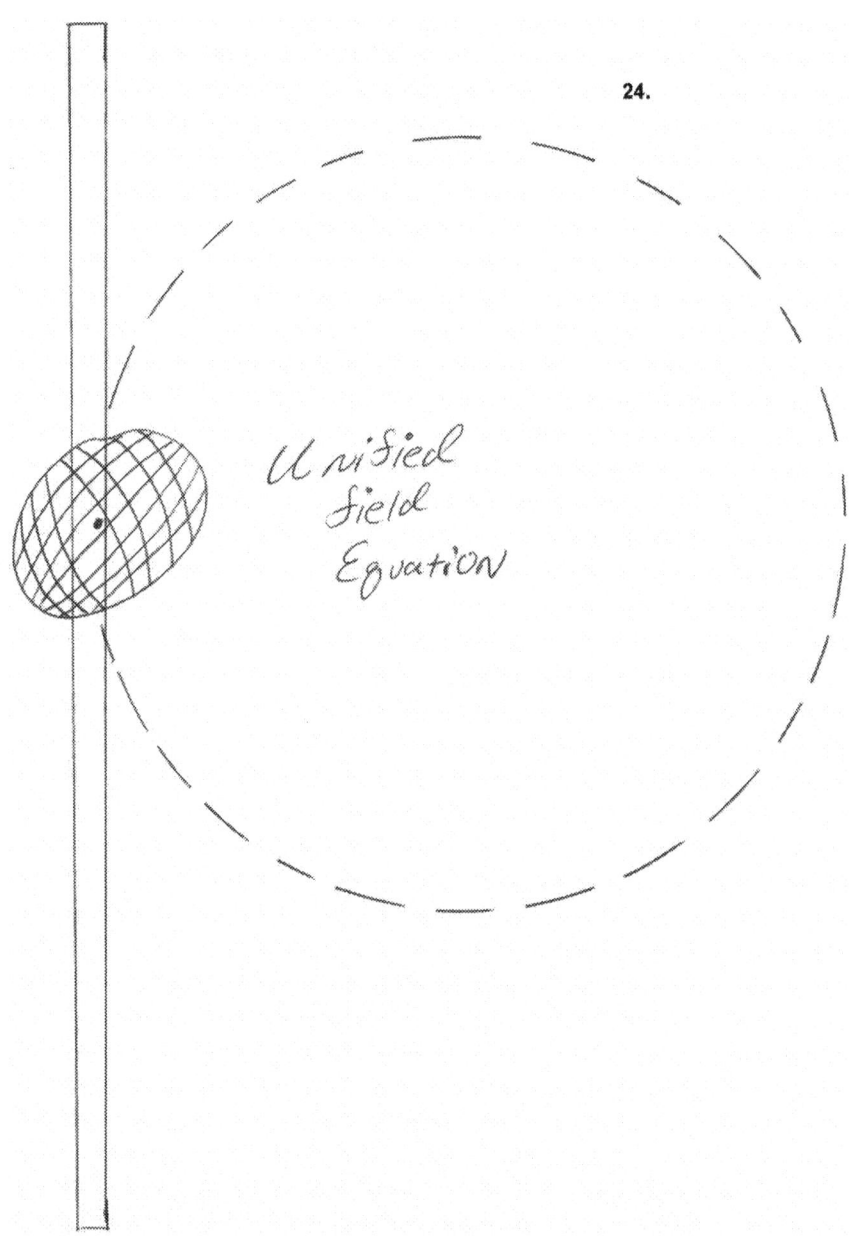

24.

Unified
field
Equation

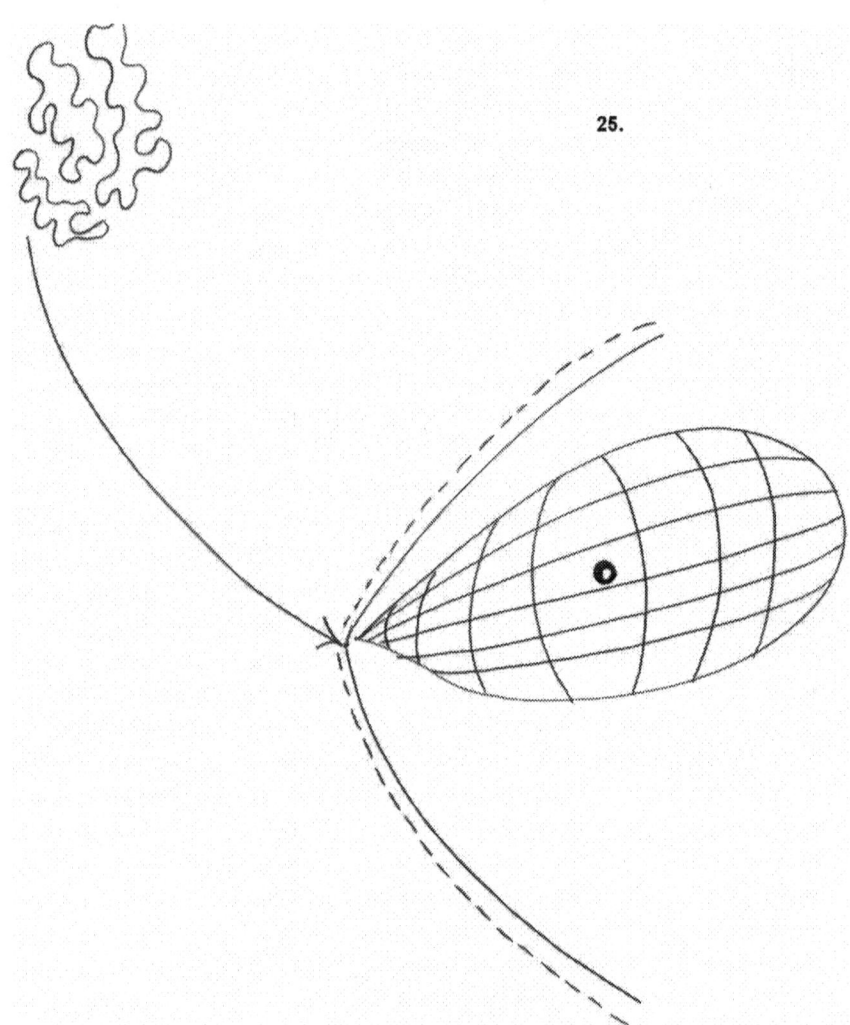

25.

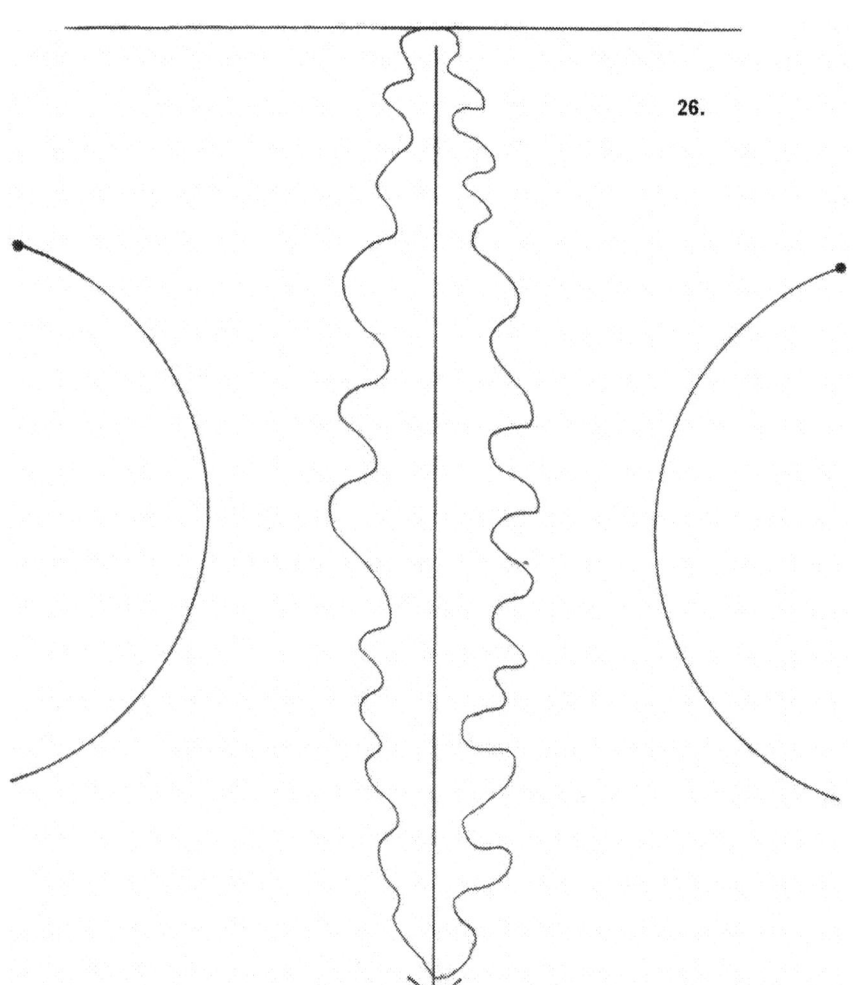

26.

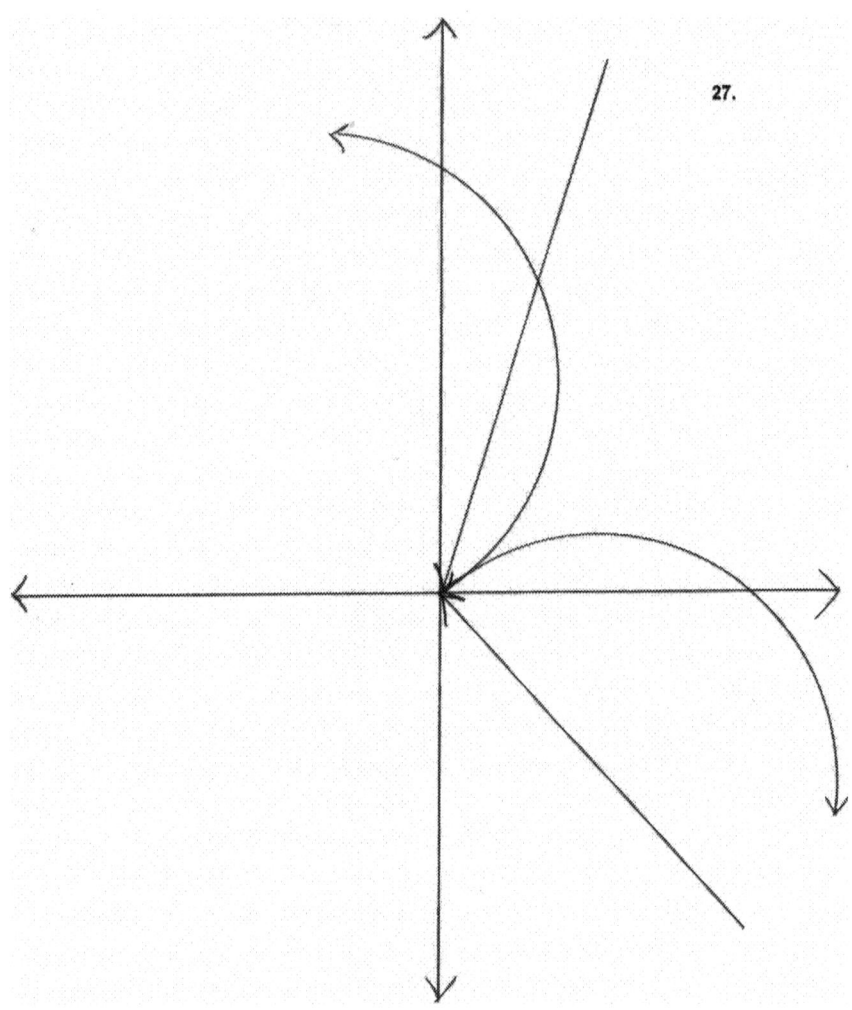

27.

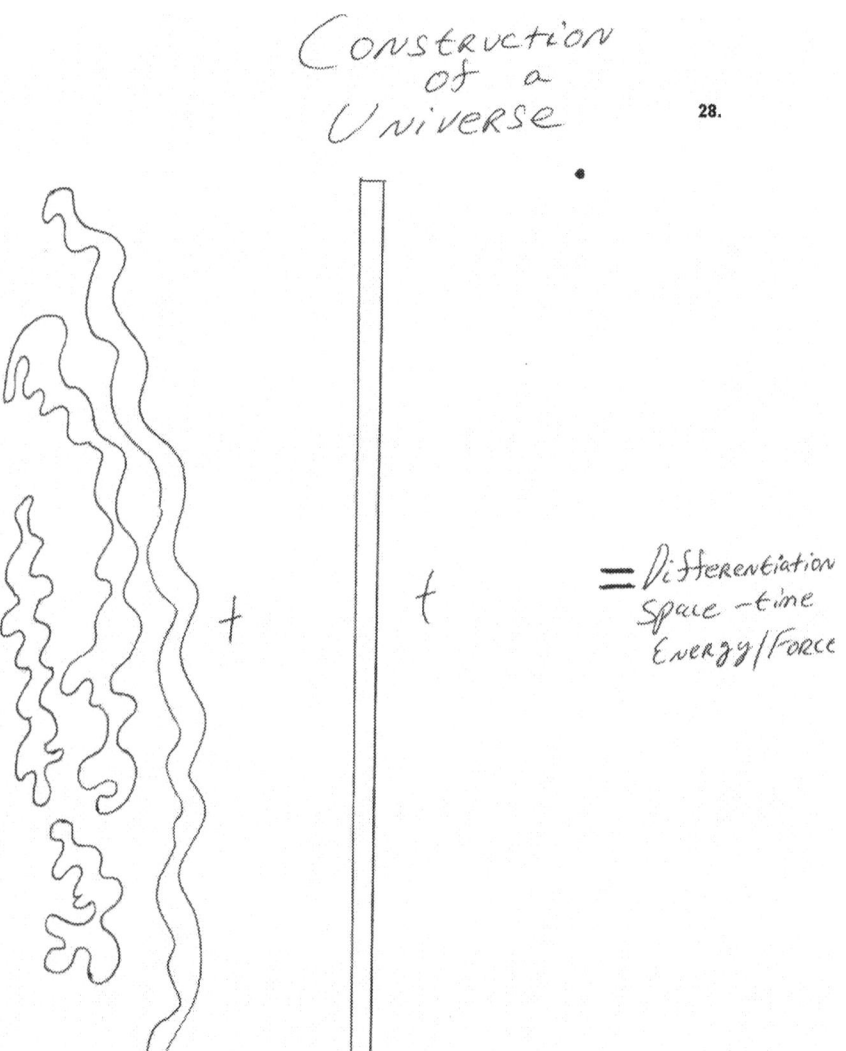

Construction
of a
Universe

28.

t t = Differentiation
Space—time
Energy/Force

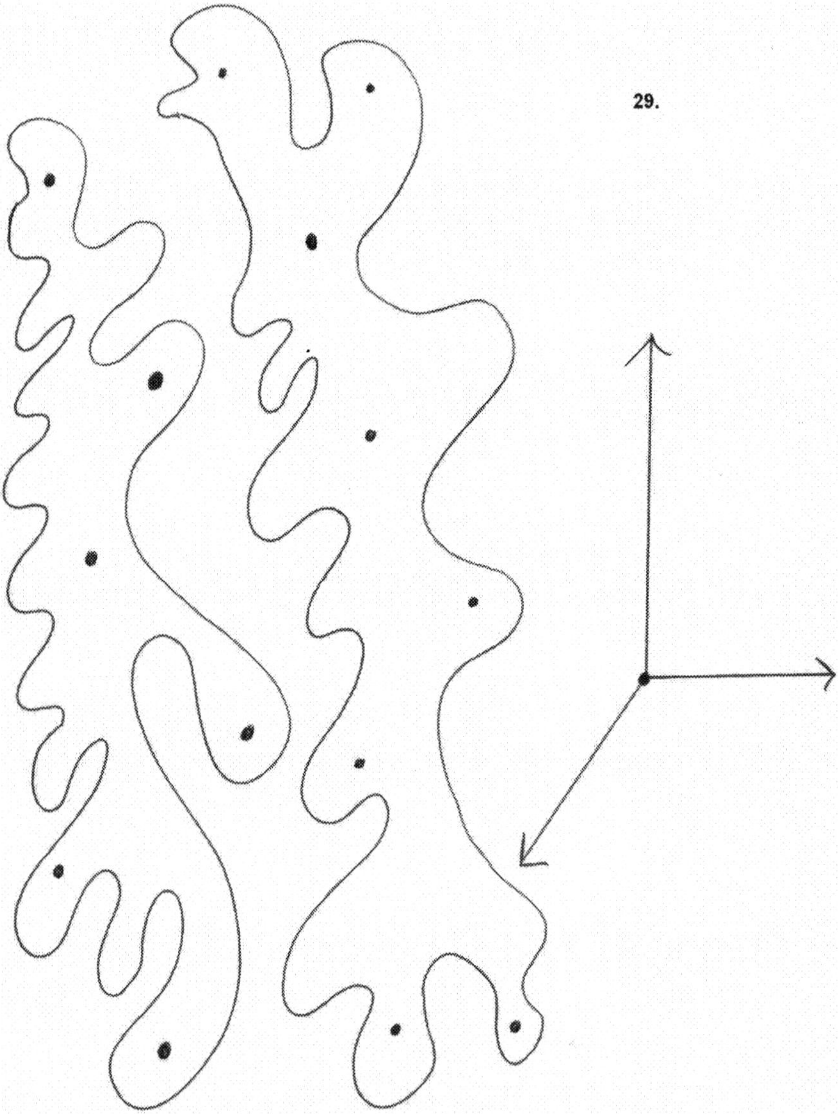

29.

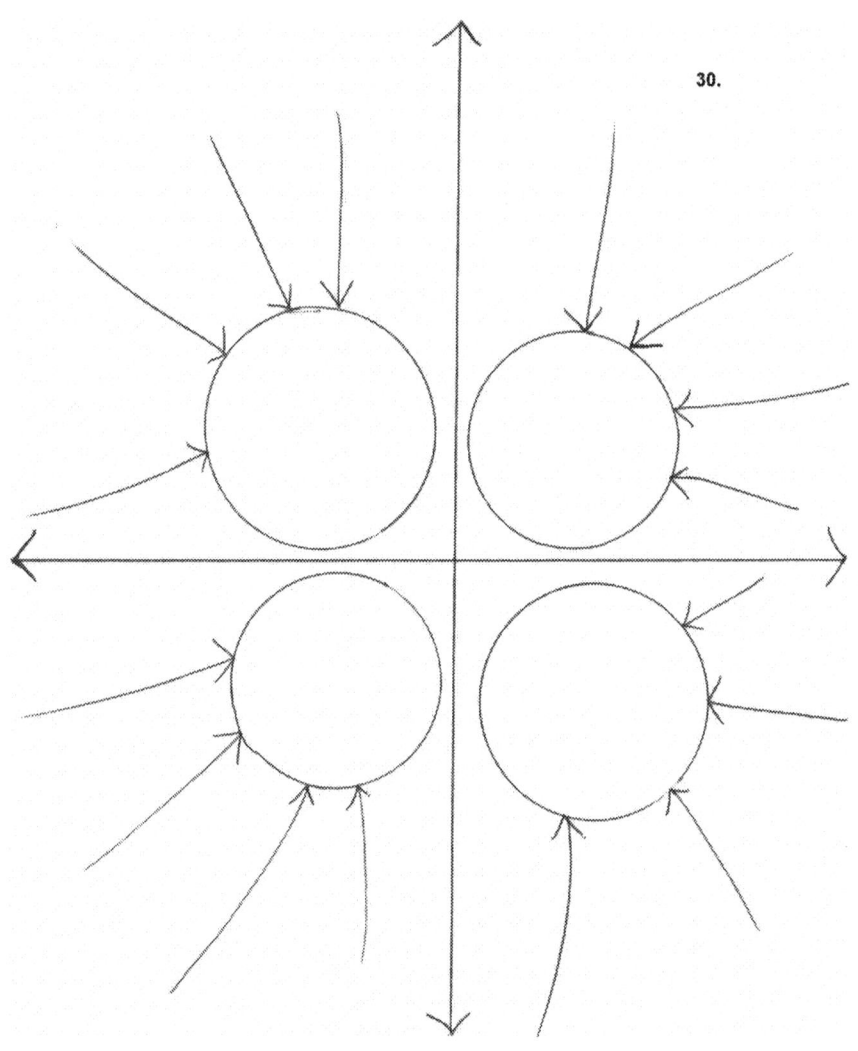

30.

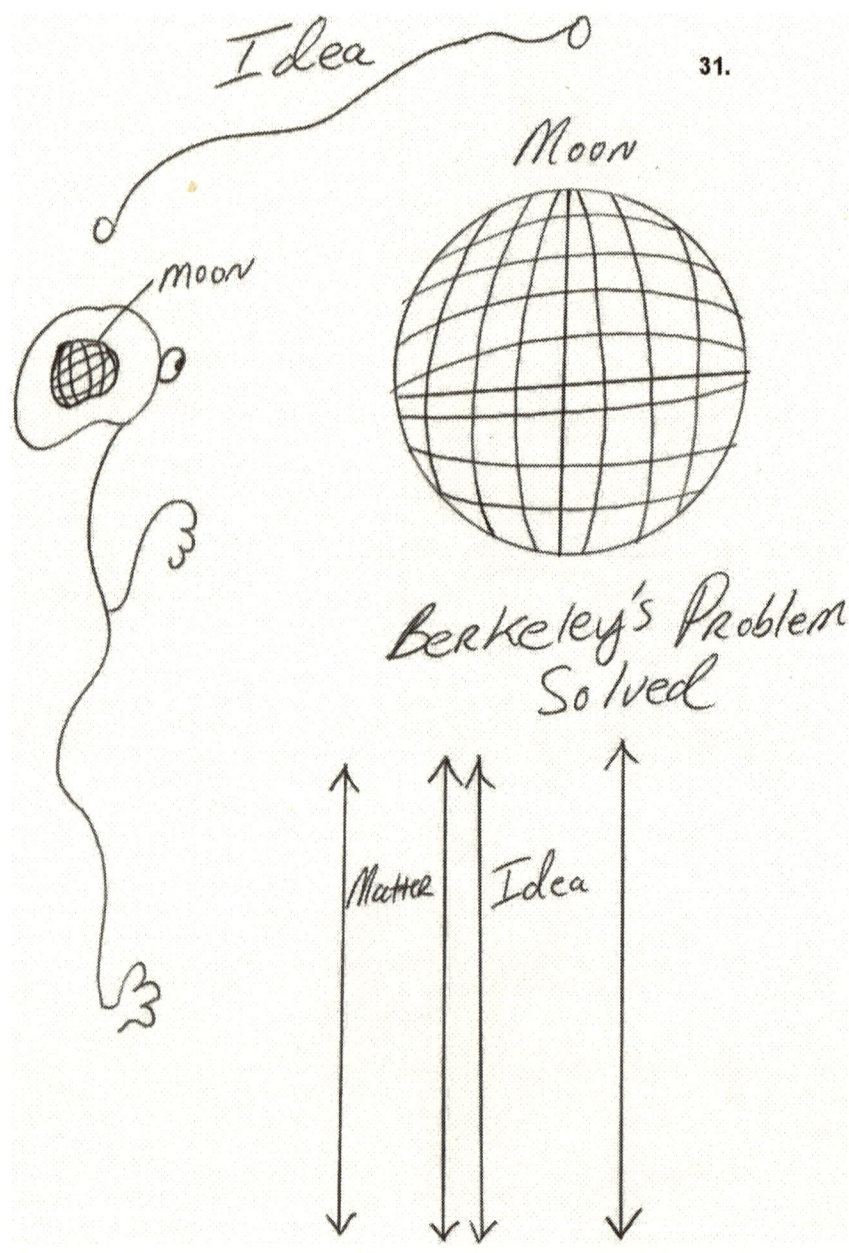

Idea

Moon

moon

31.

Moon

Berkeley's Problem
Solved

Matter Idea

32.

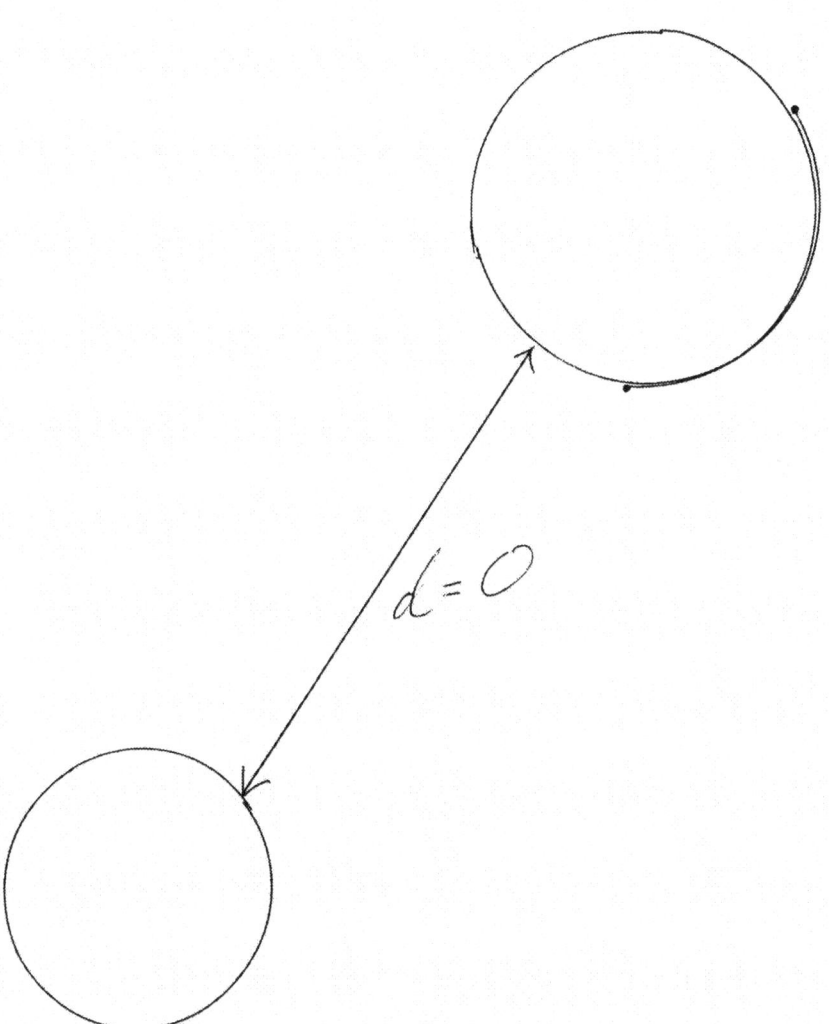

33.

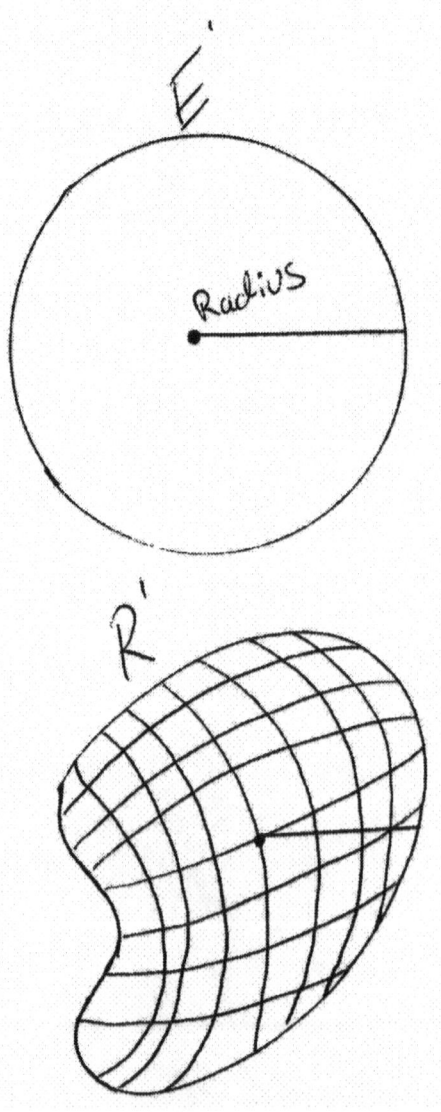

34.

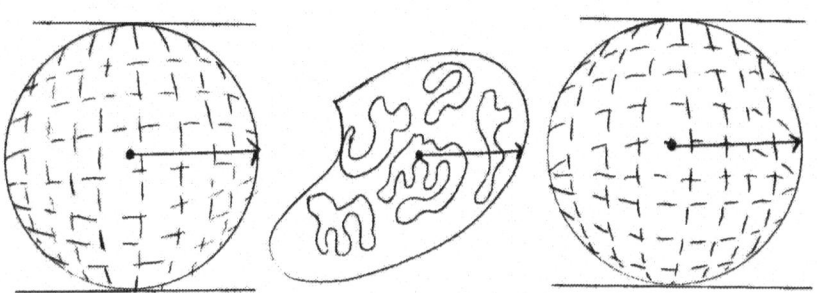

35.

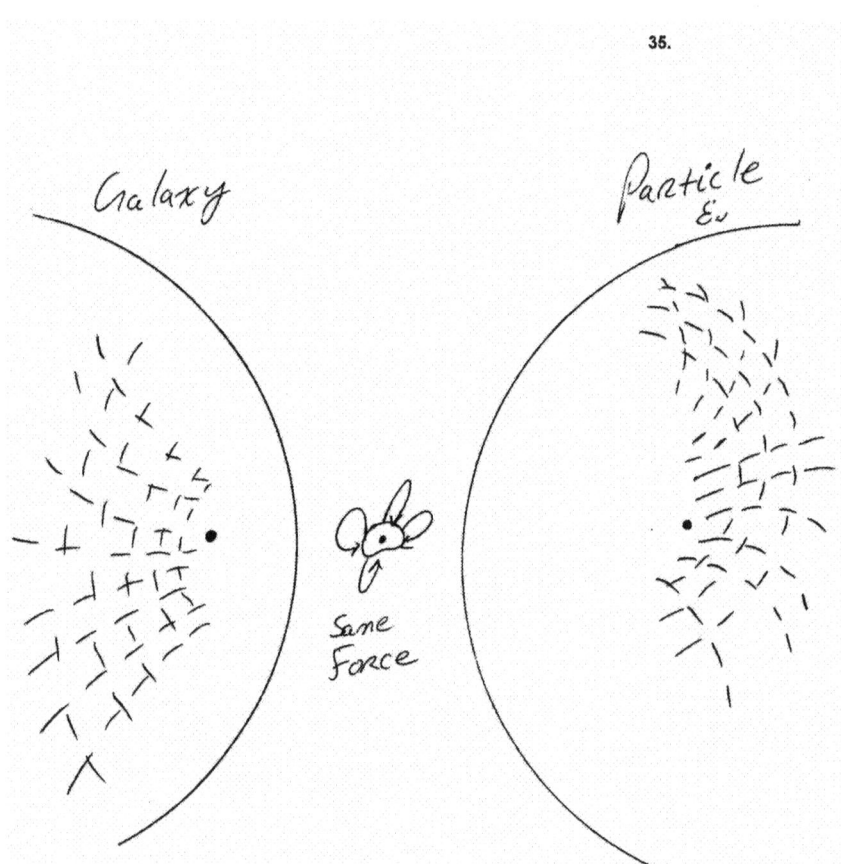

36.

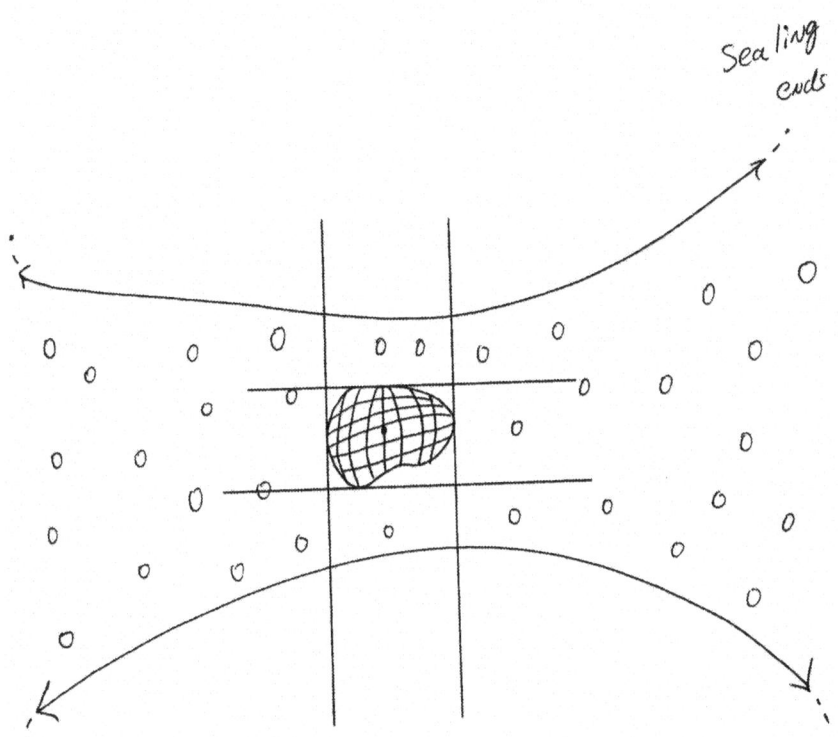

Sealing ends

37.

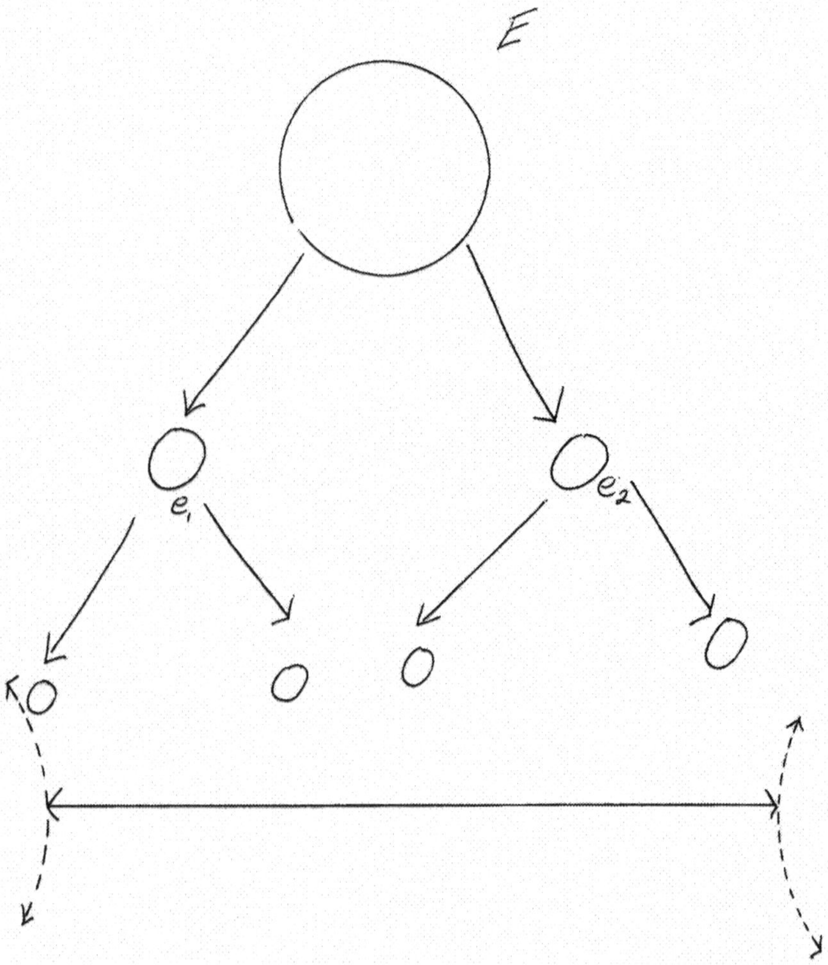

E

e_1

e_2

38.

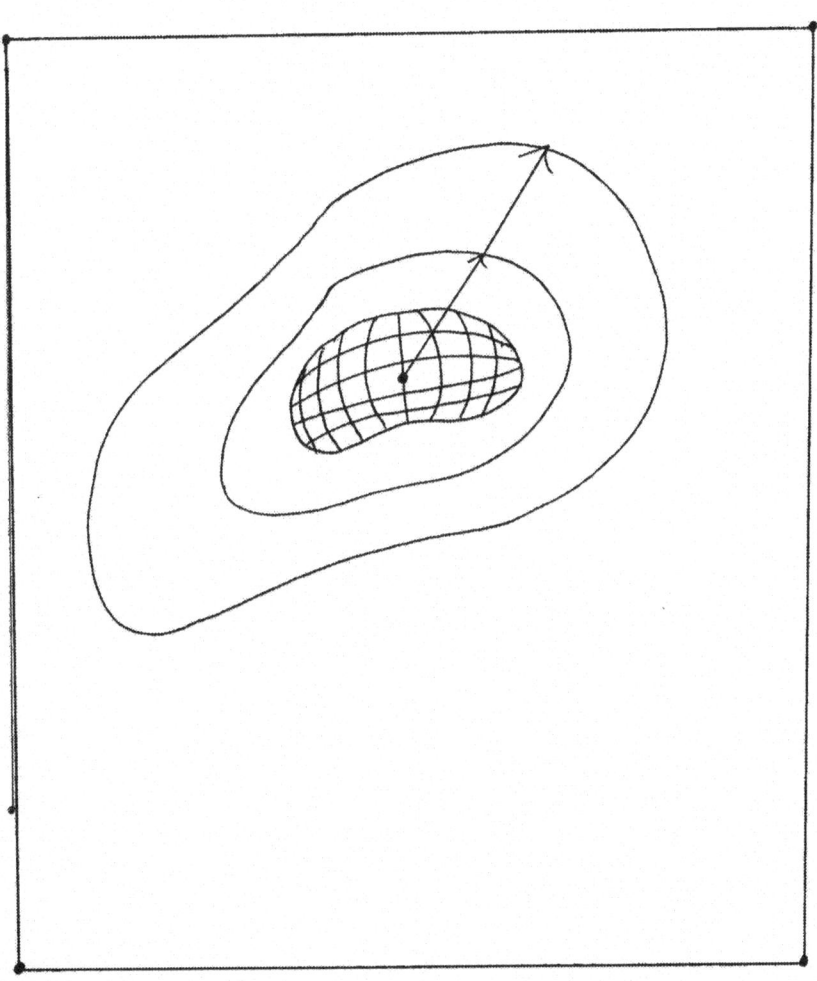

978-0-595-39017-5
0-595-39017-X

www.ingramcontent.com/pod-product-compliance
Lightning Source LLC
Chambersburg PA
CBHW021036180526
45163CB00005B/2155